Algebra 1

Applications • Equations • Graphs

Standardized Test Practice Workbook
Teacher's Edition

The Standardized Test Practice Workbook provides practice exercises for every lesson in a standardized test format. Included are multiple-choice, quantitative-comparison, and multi-step problems. The Teacher's Edition includes the student workbook and the answers.

McDougal Littell
A HOUGHTON MIFFLIN COMPANY
Evanston, Illinois • Boston • Dallas

ISBN: 0-618-02054-3

123456789-CKI- 04 03 02 01 00

Contents

NAME _____ DATE _____

Standardized Test Practice

For use with pages 3–8

TEST TAKING STRATEGY   **Spend no more than a few minutes on each question.**

1. *Multiple Choice*   What is the value of the expression $7 - x$ when $x = 2$?

Ⓐ 14   Ⓑ −5   Ⓒ −9
Ⓓ 5   Ⓔ −19

2. *Multiple Choice*   What is the value of the expression $2x + 3$ when $x = 3$?

Ⓐ 8   Ⓑ −3   Ⓒ 9
Ⓓ 3   Ⓔ −9

3. *Multiple Choice*   Evaluate $3y - 6$ when $y = 4$.

Ⓐ 6   Ⓑ −18   Ⓒ 18
Ⓓ −6   Ⓔ −12

4. *Multiple Choice*   Find the area of a triangle if the height is 6 inches and the base is 9 inches. (*Hint*: Area of a triangle is $A = \frac{1}{2}bh$.)

Ⓐ 7.5   Ⓑ 18   Ⓒ 15
Ⓓ 54   Ⓔ 27

5. *Multiple Choice*   Evaluate $\frac{6}{7}x$ when $x = \frac{2}{3}$.

Ⓐ $\frac{8}{10}$ Ⓑ $1\frac{11}{21}$ Ⓒ $\frac{12}{7}$
Ⓓ $\frac{4}{5}$ Ⓔ $\frac{4}{7}$

6. *Multiple Choice*   The variable expression for 3 times x plus 1.2 is ___?___ .

Ⓐ $1.2x + 3$ Ⓑ $3x + 3.6$
Ⓒ $3x + 1.2$ Ⓓ $x + 1.2 \cdot 3$
Ⓔ $\frac{x}{3} + 1.2$

7. *Multiple Choice*   Which is the variable expression for 26 divided by w plus 5?

Ⓐ $\frac{26}{w} + 5$ Ⓑ $\frac{w}{26} + 5$
Ⓒ $26w + 5$ Ⓓ $26 \div 5w$
Ⓔ $\frac{w}{5} - 26$

8. *Multiple Choice*   A bat can eat 0.05 pound of mosquitoes in an hour. If it eats for 6.5 hours, how many pounds of mosquitoes will it consume?

Ⓐ 0.6 lb   Ⓑ 0.325 lb
Ⓒ 0.13 lb   Ⓓ 3.25 lb   Ⓔ 1.3 lb

9. *Multiple Choice*   Your parents invested $1000 in a college savings account earning 5.5%. If this was invested 12 years ago, how much interest has it earned?

Ⓐ $66.00   Ⓑ $660.00
Ⓒ $666.00   Ⓓ $2182.00
Ⓔ $60.00

Quantitative Comparison   In Exercises 10–13, choose the statement below that is true about the given numbers.

Ⓐ The number in column A is greater.
Ⓑ The number in column B is greater.
Ⓒ The two numbers are equal.
Ⓓ The relationship cannot be determined from the given information.

	Column A	Column B
10.	$\frac{6}{x}$ when $x = 4$	$3y - 6$ when $y = 3$
11.	10 times y	10 divided by y
12.	$3z + 4$ when $z = \frac{1}{3}$	$\frac{1}{2}z - 3$ when $z = 10$
13.	$\frac{2}{3}a + \frac{1}{2}$ when $a = \frac{1}{4}$	$\frac{2}{5}a - \frac{1}{3}$ when $a = 7$

TEST TAKING STRATEGY   As soon as the testing time begins, start working. Keep moving and stay focused on the test.

1. Multiple Choice   Which algebraic expression is equivalent to $6 \cdot 6 \cdot 6 \cdot 6 \cdot 6$?

  A 6(5)   **B** 5^6 **C** 7776

  D 30   **E** 6^5

2. Multiple Choice   Express the meaning of the power 3^7 in numbers.

  A $7 \cdot 7 \cdot 7$ **B** $3 \cdot 3 \cdot 3 \cdot 3 \cdot 3 \cdot 3 \cdot 3$

  C 3(7)   **D** 2191

  E $3 \cdot 3 \cdot 3 \cdot 3 \cdot 3 \cdot 3$

3. Multiple Choice   Evaluate the expression x^5 when $x = 3$.

  A 125   **B** 243   **C** 15

  D 8   **E** 81

4. Multiple Choice   Evaluate the expression $x^2 - 3y$ when $x = 5$ and $y = 3$.

  A 16   **B** 19   **C** 1

  D −6   **E** 34

5. Multiple Choice   Evaluate the expression $(m + p)^3$ when $m = 4$ and $p = 3$.

  A 67   **B** 31   **C** 21

  D 1728   **E** 343

6. Multiple Choice   Evaluate the expression $2x + (y^2)$ when $x = 7$ and $y = 4$.

  A 25   **B** 17   **C** 144

  D 30   **E** 22

7. Multiple Choice   Evaluate the expression $(3a)^4$ when $a = 2$.

  A 625   **B** 162   **C** 48

  D 1296   **E** 24

8. Multiple Choice   Halie's bedroom is a square measuring 10 feet along the base of each wall and 10 feet high. Which expression is used to calculate the square footage needed to wallpaper her room?

  A 10^4 **B** 10^3

  C $4(10^2)$ **D** $4^2 \cdot 10$

  E $(4 \cdot 10)^2$

9. Multiple Choice   A cylindrical vase is 14 inches high and has a radius of 5 inches. Determine the volume if the formula is 3.14 times the square of the radius times height.

  A 3077   **B** 353.14   **C** 1099

  D 439.6   **E** 1380

10. Multi-Step Problem   You have a large rectangular fish tank measuring 3 feet across by 2 feet deep by 2 feet high. You want to fill it $\frac{3}{4}$ of the way with water.

  a. What is the volume of the fish tank in cubic feet?

  b. If one cubic foot is equal to 7.48 gallons, how many gallons of water can the tank hold?

  c. How many gallons will fill the tank $\frac{3}{4}$ full?

  d. Would a fish tank measuring 3 feet across by 4 feet deep by 2 feet high double the volume of your current fish tank?

Chapter 1

NAME _____ DATE _____

Standardized Test Practice

For use with pages 16–22

TEST TAKING STRATEGY   **If you can, check an answer using a method that is different from the one you used originally, to avoid making the same mistake twice.**

1. *Multiple Choice*   Evaluate the expression $4x^2 - 2$ when $x = 3$.

   Ⓐ 47   Ⓑ 142   Ⓒ 22

   Ⓓ 72   Ⓔ 34

2. *Multiple Choice*   Evaluate the expression $126 \div 2y^3$ when $y = 3$.

   Ⓐ $\dfrac{7}{12}$ Ⓑ 1701   Ⓒ 7

   Ⓓ $2\dfrac{1}{3}$ Ⓔ 567

3. *Multiple Choice*   Evaluate the expression $\dfrac{88}{x^4} + 6x$ when $x = 2$.

   Ⓐ 23   Ⓑ 3.14   Ⓒ 17.5

   Ⓓ 4.4   Ⓔ 13.5

4. *Multiple Choice*   Evaluate the expression $\frac{2}{3}x^2 - \frac{1}{12}$ when $x = \frac{1}{2}$.

   Ⓐ $\dfrac{1}{12}$ Ⓑ $\dfrac{29}{84}$ Ⓒ $\dfrac{1}{4}$

   Ⓓ $\dfrac{1}{3}$ Ⓔ $\dfrac{7}{12}$

5. *Multiple Choice*   Evaluate the expression $6 \cdot (3^2 + 6)$.

   Ⓐ 72   Ⓑ 66   Ⓒ 60

   Ⓓ 36   Ⓔ 90

6. *Multiple Choice*   Insert grouping symbols into $1 + 4 \cdot 3^2 + 2$ to produce the value 45.

   Ⓐ $1 + (4 \cdot 3)^2 + 2$ Ⓑ $1 + 4 \cdot (3^2 + 2)$

   Ⓒ $1 + 4 \cdot (3^2) + 2$ Ⓓ $(1 + 4) \cdot 3^2 + 2$

   Ⓔ $(1 + 4 \cdot 3)^2 + 2$

7. *Multiple Choice*   Evaluate $\frac{1}{3}(48 + 3) - 2^3$.

   Ⓐ 41   Ⓑ 9   Ⓒ 24

   Ⓓ 40   Ⓔ 11

8. *Multiple Choice*   What is the value of $\dfrac{15 - 6 \cdot 2}{8 \cdot (4^2 - 7)}$?

   Ⓐ $2\dfrac{1}{4}$ Ⓑ $\dfrac{1}{24}$ Ⓒ $\dfrac{3}{8}$

   Ⓓ $\dfrac{3}{121}$ Ⓔ $\dfrac{3}{4}$

9. *Multiple Choice*   If your state charges a 6% sales tax on all items except food and clothing, what is the sales tax if you bought groceries for $22.15, a pair of jeans for $32.00, a video for $16.00, and books for $35.00?

   Ⓐ $6.31   Ⓑ $4.39   Ⓒ $4.98

   Ⓓ $2.10   Ⓔ $3.06

Quantitative Comparison   In Exercises 10–12, choose the statement below that is true about the given numbers.

   Ⓐ The number in column A is greater.

   Ⓑ The number in column B is greater.

   Ⓒ The two numbers are equal.

   Ⓓ The relationship cannot be determined from the given information.

	Column A	Column B
10.	$(19 - 4)^2 + 5$	$19 - 4^2 + 5$
11.	$18 + 9 \div 3 - 6$	$15 \div 3 \cdot 2^2 - 5$
12.	$3 + 6^2 \div 4 + 2$	$12 \div 3 \cdot 2^2 - 1$

Chapter 1

Standardized Test Practice

For use with pages 24–30

TEST TAKING STRATEGY   Avoid spending too much time on one question.

1. Multiple Choice   The solution of the equation $7x + 2 = 23$ is __?__ .

  Ⓐ 4 Ⓑ 3 Ⓒ 14 Ⓓ 12 Ⓔ 5

2. Multiple Choice   The solution of the equation $15 - 3y = 7$ is __?__ .

  Ⓐ 5 Ⓑ 4 Ⓒ 2 Ⓓ 3 Ⓔ 6

3. Multiple Choice   The solution of the equation $10x - 6 = 10 + 6x$ is __?__ .

  Ⓐ 1 Ⓑ 4 Ⓒ 3 Ⓓ 5 Ⓔ 2

4. Multiple Choice   Determine which equation has a solution of $x = 3$.

  Ⓐ $4x + 6 = 18$ Ⓑ $16 = 12x + 1$

  Ⓒ $18 - 3x = 12$ Ⓓ $4 + 2x = 3x - 1$

  Ⓔ $5 + 2x = 13$

5. Multiple Choice   Determine which equation has a solution of $a = 5$.

  Ⓐ $2a + 6 = 10$ Ⓑ $32 = 7a + 4$

  Ⓒ $12 - a = 4$ Ⓓ $25 - 3a = 10$

  Ⓔ $16 - a^2 = 8$

6. Multiple Choice   Which equation does not have a solution of $w = 6$?

  Ⓐ $3w - 4 = 14$ Ⓑ $10w - 22 = 38$

  Ⓒ $12 + 2w = 18$ Ⓓ $w^2 + 3 = 39$

  Ⓔ $16 = 22 - w$

7. Multiple Choice   Which equation is a translation of "3 times a number decreased by 6 gives 18"?

  Ⓐ $3x \div 6 = 18$ Ⓑ $6 - 3x = 18$

  Ⓒ $6 - 3x = 18$ Ⓓ $3x + 6 = 18$

  Ⓔ $3x - 6 = 18$

8. Multiple Choice   Using the interest formula, $I = Prt$, determine the amount of principle that must be invested to earn \$50 interest over 2 years at a rate of 5%.

  Ⓐ \$500   Ⓑ \$200   Ⓒ \$50

  Ⓓ \$2000   Ⓔ \$5000

9. Multiple Choice   Determine which inequality has a possible solution of $x = 3$.

  Ⓐ $2x < 8$ Ⓑ Ⓐ and Ⓑ

  Ⓒ $10 - 2x < 3$ Ⓓ Ⓐ and Ⓔ

  Ⓔ $15 > 4x - 6$

10. Multiple Choice   Match the sentence "the sum of 10 and twice x is less than 25" with its mathematical representation.

  Ⓐ $10(2 + x) < 25$ Ⓑ $10 + 2x < 25$

  Ⓒ $10 + 2x > 25$ Ⓓ $2 + 10x < 25$

  Ⓔ $10 - 2x < 25$

11. Multi-Step Problem   You are saving money to go to summer camp. You need at least \$500 to cover all your costs, and you've already saved \$150. You can save an additional \$20 every week.

  a. Write an inequality to model the situation using w for weeks.

  b. What do the 150 and the 20 represent?

  c. How long will it take you to save enough money?

  d. If you only save \$15 a week, how long will it take you to save enough money?

NAME _____ DATE _____

Standardized Test Practice

For use with pages 32–39

TEST TAKING STRATEGY   **Skip questions that are too difficult for you, and spend no more than a few minutes on each question.**

1. *Multiple Choice*   Choose the correct algebraic translation of "five less than the sum of a number and six."

   (A) $5 - x + 6$ (B) $5 - (x + 6)$

   (C) $x - 5 + 6$ (D) $(x + 6) - 5$

   (E) $(5 - x) + 6$

2. *Multiple Choice*   Choose the correct algebraic translation of "five plus the product of eight and a number."

   (A) $8(5 + x)$ (B) $5 + 8x$ (C) $5 + \dfrac{8}{x}$

   (D) $8 + 5x$ (E) $5 + (8 + x)$

3. *Multiple Choice*   Choose the correct algebraic translation of "three squared plus the quantity of six times a number."

   (A) $(3^2 + 6)y$ (B) $3^2 + 6y$

   (C) $3(2) + 6y$ (D) $3^2 + \dfrac{6}{y}$

   (E) $3(2) + \dfrac{6}{y}$

4. *Multiple Choice*   Choose the correct algebraic translation of "the product of 12 and a number is 7."

   (A) $12x = 7$ (B) $\dfrac{x}{12} = 7$

   (C) $12 + x = 7$ (D) $12 = 7x$

   (E) $\dfrac{12}{x} = 7$

5. *Multiple Choice*   Choose the correct algebraic translation of "thirty-six is the product of w and 12."

   (A) $36 = w + 12$ (B) $36 = \dfrac{w}{12}$

   (C) $w - 12 = 36$ (D) $36w = 12$

   (E) $12w = 36$

6. *Multiple Choice*   Choose the correct algebraic translation of "seven less than the product of five and a number is less than fourteen."

   (A) $7 - 5x < 14$ (B) $7 - 5x > 14$

   (C) $5x - 7 < 14$ (D) $5x - 7 > 14$

   (E) $7 < 5x + 14$

7. *Multiple Choice*   The lease agreement on a new car requires a downpayment of $1200 and $269 a month for 24 months. There is an additional fee of $.15 per mile for all miles, m, driven over 25,000 miles. Choose the correct model for the cost, C, of the car, if you drive more than 25,000 miles.

   (A) $1200 + 269 + 0.15(25,000 - m) = C$

   (B) $1200 + 269(24) + 0.15(25,000 - m) = C$

   (C) $1200 + 269 + 24 + 0.15m = C$

   (D) $1200 + 269(24) + 0.15m = C$

   (E) $1200 + 269(24) + 0.15(m - 25,000) = C$

Quantitative Comparison   In Exercises 8–11, choose the statement below that is true about the unknown numbers.

   (A) The unknown in column A is greater.

   (B) The unknown in column B is greater.

   (C) The two unknowns are equal.

   (D) The relationship cannot be determined from the given information.

	Column A	Column B
8.	A number decreased by three squared is six.	Ten times a number is five squared times six.
9.	A number divided by five is less than eight.	The product of 14 and a number is greater than 11.
10.	The product of a number and eight is 32.	The quotient of a number and sixteen is five.
11.	Seven times a number decreased by six is 15.	Nine more than sixteen divided by a number is seventeen.

Standardized Test Practice
For use with pages 40–45

TEST TAKING STRATEGY **Work as fast as you can through the easier problems, but not so fast that you are careless.**

Multiple Choice In Exercises 1–4, use the following information and graph to determine the correct answer.

Australia has the highest incidence of pet ownership in the world, with 66% of households owning a pet.

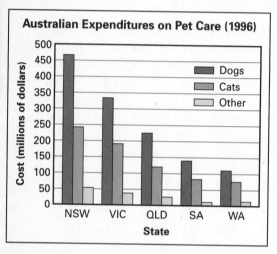

Australian Expenditures on Pet Care (1996)

1. Which state has the highest total spending on pet care?

   (A) NSW   (B) VIC   (C) QLD

   (D) SA   (E) WA

2. Which two states spent about the same amount of money on "other" pet care?

   (A) NSW and VIC   (B) VIC and QLD

   (C) QLD and SA   (D) SA and WA

   (E) WA and QLD

3. What is the approximate total expenditure on dog and cat care in SA?

   (A) $345 million   (B) $710 million

   (C) $220 million   (D) $525 million

   (E) $150 million

4. Which state spent approximately $150 million on cat and other pet expenditures?

   (A) NSW   (B) VIC   (C) QLD

   (D) SA   (E) WA

5. *Multiple Choice* In which three years did home ownership increase?

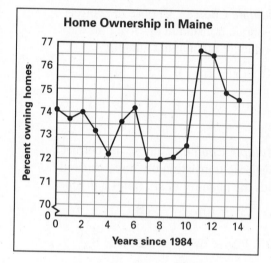

Home Ownership in Maine

   (A) 1984, 1985, 1986   (B) 1987, 1988, 1989

   (C) 1996, 1997, 1998   (D) 1986, 1987, 1988

   (E) 1993, 1994, 1995

Quantitative Comparison In Exercises 6–8, refer to the graph in exercise 5 and choose the statement that is true about the given numbers.

   (A) The number in column A is greater.

   (B) The number in column B is greater.

   (C) The two numbers are equal.

   (D) The relationship cannot be determined from the given information.

	Column A	Column B
6.	ownership rate in 1987	ownership rate in 1990
7.	ownership rate in 1990	ownership rate in 1994
8.	ownership rate in 1991	ownership rate in 1992

Algebra 1
Standardized Test Practice Workbook

Chapter 7

Standardized Test Practice

For use with pages 46–52

Chapter 1

TEST TAKING STRATEGY   **If you can, check your answer using a method that is different from the one you used originally, to avoid making the same mistake twice.**

1. *Multiple Choice*   Choose the correct domain for the function $t = 35 + 5w$ where $0 \le w \le 5$ and w is an integer.

  Ⓐ 0, 1, 2, 3, 4, 5

  Ⓑ 1, 2, 3, 4, 5

  Ⓒ 35, 40, 45, 50, 55

  Ⓓ 40, 45, 50, 55

  Ⓔ 40, 41, 42, 43, 44, 45

2. *Multiple Choice*   Choose the correct range for the function $y = 250 + 10x$ where $0 \le x \le 4$ and x is an integer.

  Ⓐ 0, 1, 2, 3, 4

  Ⓑ 0, 1, 2, 3

  Ⓒ 250, 260, 270, 280, 290

  Ⓓ 250, 260, 270, 280

  Ⓔ 260, 261, 262, 263, 264

3. *Multiple Choice*   Choose the table that does not represent a function.

Ⓐ

Input	0	1	2	3	4
Output	10	15	20	25	30

Ⓑ

Input	1	2	3	4	5
Output	3	3	5	5	7

Ⓒ

Input	0	0	1	1	2
Output	7	9	11	13	15

Ⓓ

Input	0	1	2	3	4
Output	5	10	15	30	90

Ⓔ

Input	1	2	3	4	5
Output	6	12	18	24	30

4. *Multiple Choice*   Choose the equation that describes the function containing all of the points shown in the table.

Input	0	1	2	3	4	5
Output	7	15	23	31	39	47

  Ⓐ $y = 12 + x^2$ Ⓑ $y = 7 + 2x^2$

  Ⓒ $y = 4x + 7$ Ⓓ $y = 16 - 2x$

  Ⓔ $y = 8x - 7$

5. *Multiple Choice*   A bag of puppy food directs you to feed $\frac{1}{2}$ cup for every 10 pounds the puppy weighs. Choose the correct representation of cups, c, as a function of body weight, w.

  Ⓐ $c = \frac{1}{2}(10w)$ Ⓑ $c = \frac{1}{2}\left(\frac{w}{10}\right)$

  Ⓒ $c = \frac{1}{2} + w$ Ⓓ $w = \frac{1}{2} + 10c$

  Ⓔ $c = \frac{1}{2}(w - 10)$

6. *Multi-Step Problem*   Your high school band is selling sub sandwiches for a fund raiser. It costs $1.25 to make each sandwich. They sell them for $2.50 each.

  a. Write a function that determines the profit from the sale.

  b. The band must pay a $75 fee to use the cafeteria to assemble the sandwiches. Write a function that includes this cost.

  c. The band wants $1000 profit. How many sandwiches must they sell?

  d. If they sell 820 sandwiches, what was the profit earned?

NAME _____ DATE _____

Standardized Test Practice

For use with pages 63–70

TEST TAKING STRATEGY **If you can, check an answer using a method that is differ-ent from the one you used originally, to avoid making the same mistake twice.**

1. *Multiple Choice* Choose the number repre-sented on the graph.

  A $\frac{1}{2}$ **B** $-\frac{1}{2}$ **C** $\frac{1}{4}$

  D $-\frac{1}{4}$ **E** -1

2. *Multiple Choice* Choose the correct ordering (in increasing order) of the following numbers: $2, -0.5, 0, -1, 0.75$

  A $0, -0.5, .75, -1, 2$

  B $-0.5, -1, 0, 0.75, 2$

  C $-0.5, -1, 0.75, 0, 2$

  D $-1, -0.5, 0, 0.75, 2$

  E $-1, -0.5, 0.75, 0, 2$

3. *Multiple Choice* Choose the correct ordering (in increasing order) of the following numbers: $0, 6, -3, \frac{1}{2}, -0.5, -\frac{3}{4}$

  A $0, -3, -0.5, -3, \frac{1}{2}, -0.5, -\frac{3}{4}$

  B $-3, -0.5, -\frac{3}{4}, 0, \frac{1}{2}, 6$

  C $0, -0.5, \frac{1}{2}, -\frac{3}{4}, -3, 6$

  D $-3, -\frac{3}{4}, -0.5, \frac{1}{2}, 0, 6$

  E $-3, -\frac{3}{4}, -0.5, 0, \frac{1}{2}, 6$

Multiple Choice For Exercises 4 and 5, refer to the table showing high temperatures in January for one week in Pittsburgh, PA.

Date	$\frac{1}{3}$	$\frac{1}{4}$	$\frac{1}{5}$	$\frac{1}{6}$	$\frac{1}{7}$	$\frac{1}{8}$	$\frac{1}{9}$
°C	$10°$	$5°$	$-5°$	$-10°$	$3°$	$0°$	$-2°$

4. What was the coldest temperature recorded?

  A $0°$ **B** $3°$ **C** $-2°$

  D $-5°$ **E** $-10°$

5. Which days had high temperatures above freezing (0°C)?

  A Jan. 3, 4, 7   **B** Jan. 3, 4, 7, 8

  C Jan. 3, 4, 5, 6, 7, 9   **D** Jan. 3, 4

  E Jan. 5, 6, 9

6. *Multiple Choice* Choose the inequality that would correctly compare -4 and -6.

  A $-4 = -6$ **B** $-4 < -6$

  C $-6 < -4$ **D** $-6 > -4$

  E Both B and C

7. *Multiple Choice* Evaluate the expression $-|-2|$.

  A -2 **B** 2 **C** 0

  D 1 **E** -1

8 *Multiple Choice* Evaluate the expression $|-7| + 3$.

  A -4 **B** 4 **C** 3

  D 10 **E** -10

9. *Multiple Choice* If $-6 \geq w$, then $w \underline{\ ?\ } -6$.

  A $<$ **B** $>$ **C** $\leq$

  D $\geq$ **E** $=$

Quantitative Comparison In Exercises 10–12, choose the statement that is true about the given numbers.

  A The number in column A is greater

  B The number in column B is greater

  C The two numbers are equal

  D The relationship cannot be determined from the information given.

	Column A	Column B
10.	$-|3 + 2|$	$|3 + 2|$
11.	$|x| + 2$	$|-x| + 2$
12.	$|x - 6|$	$|x| - 6$

Algebra 1
Standardized Test Practice Workbook

NAME _____ DATE _____

Standardized Test Practice
For use with pages 72–77

TEST TAKING STRATEGY   Spend no more than a few minutes on each question.

1. *Multiple Choice*   Find the sum of $3 + (-4)$.
   - Ⓐ 7
   - Ⓑ −7
   - Ⓒ −1
   - Ⓓ 1
   - Ⓔ 12

2. *Multiple Choice*   Find the sum of $-6 + (-18)$.
   - Ⓐ −12
   - Ⓑ 12
   - Ⓒ 24
   - Ⓓ −24
   - Ⓔ 3

3. *Multiple Choice*   Find the sum of $-5 + 8 + (-2)$.
   - Ⓐ 15
   - Ⓑ −15
   - Ⓒ 1
   - Ⓓ −1
   - Ⓔ 11

4. *Multiple Choice*   Find the sum of $2.3 + (-0.5) + (-1.2) + (-0.7)$.
   - Ⓐ −0.1
   - Ⓑ 0.1
   - Ⓒ 0.98
   - Ⓓ 4.7
   - Ⓔ −4.7

5. *Multiple Choice*   Evaluate the expression $5 + x + (-3)$ for $x = -2$.
   - Ⓐ 6
   - Ⓑ −6
   - Ⓒ 0
   - Ⓓ 1
   - Ⓔ −1

Multiple Choice   In Exercises 6–8, use the table, which shows the average daily high temperature and the actual temperature of a city in degrees celsius.

Date	Avg. High	Act. High
Jan. 8	2	4
Jan. 9	0	5
Jan. 10	−1	−3
Jan. 11	−3	−11
Jan. 12	1	−8
Jan. 13	3	−2
Jan. 14	5	0

6. On which date was the actual high temperature furthest from the average temperature?
   - Ⓐ Jan. 11
   - Ⓑ Jan. 13
   - Ⓒ Jan. 8
   - Ⓓ Jan. 12
   - Ⓔ Jan. 9

7. On which dates were the actual temperatures 5 degrees different from the average high temperatures?
   - Ⓐ Jan. 9
   - Ⓑ Jan. 9, 13
   - Ⓒ Jan. 9, 14
   - Ⓓ Jan. 13, 14
   - Ⓔ Jan. 9, 13, 14

8. On which date was the actual temperature closest to the average temperature?
   - Ⓐ Jan. 8
   - Ⓑ Jan. 10
   - Ⓒ Jan. 13
   - Ⓓ Jan. 8, 10
   - Ⓔ Jan. 8, 10, 13

Quantitative Comparison   In Exercises 9–12, choose the statement that is the true about the given quantities.
   - Ⓐ   The quantity in column A is greater.
   - Ⓑ   The quantity in column B is greater.
   - Ⓒ   The two quantities are equal.
   - Ⓓ   The relationship cannot be determined from the information given.

	Column A	Column B
9.	$-3 + (-2)$	$|3 + 2|$
10.	$-\frac{1}{2} + 1\frac{2}{3}$	$|-x| + 2$
11.	$8.2 + (-4.6) + 0.4$	$-5.1 + 7.9 + (-2)$
12.	$(-2) + (-3) + (-7)$	$8 + (-11) + (-9)$

Standardized Test Practice

For use with pages 79–85

TEST TAKING STRATEGY   Work as fast as you can through the easier problems, but not so fast that you are careless.

1. *Multiple Choice*   Find the difference of $12 - 18$.

Ⓐ 6   Ⓑ -6 Ⓒ 30

Ⓓ -30 Ⓔ $\frac{12}{18}$

2. *Multiple Choice*   Evaluate the expression $-7 - 10 + 2$.

Ⓐ 5   Ⓑ 15   Ⓒ -15

Ⓓ -19 Ⓔ 1

3. *Multiple Choice*   Evaluate the expression $-6 - (-12) + 8$.

Ⓐ 26   Ⓑ 2   Ⓒ -2

Ⓓ 14   Ⓔ -10

4. *Multiple Choice*   Find the difference of $-\frac{1}{2} - \frac{2}{3} - \left(-\frac{5}{8}\right)$.

Ⓐ $\frac{13}{24}$ Ⓑ $-\frac{13}{24}$ Ⓒ $\frac{2}{3}$

Ⓓ $-\frac{19}{24}$ Ⓔ $-1\frac{19}{24}$

5. *Multiple Choice*   Choose the correct terms of the expression $-5x - 6$.

Ⓐ -5 Ⓑ -6

Ⓒ $-5, -6$ Ⓓ $-5x, -6$

Ⓔ $-5x, 6$

6. *Multiple Choice*   Choose the correct terms of the expression $9 - 15x + 12y$.

Ⓐ $9, -15, 12$ Ⓑ $9, 15, 12$

Ⓒ $9, 15x, 12y$ Ⓓ $9, -15x, 12y$

Ⓔ $9, x, y$

7. *Multiple Choice*   Choose the outputs of the function $y = -2x + 5$ for these values of x: $-2, -1, 0,$ and 1.

Ⓐ $9, 7, 5, 3$ Ⓑ $1, 2, 3, 4$

Ⓒ $1, 3, 5, 7$ Ⓓ $1, 3, 5, 3$

Ⓔ $5, 4, 3, 4$

Multiple Choice   For Exercises 8 and 9, use the table, which shows the average price of regular unleaded gasoline in Cleveland, Ohio for five weeks.

Week	1	2	3	4	5
dollars/gallon	$1.17	$1.29	$1.09	$1.13	$1.07

8. What is the change in the price between week 1 and week 3?

Ⓐ $-\$.20$ Ⓑ $\$.20$ Ⓒ $\$.08$

Ⓓ $\$.09$ Ⓔ $-\$.08$

9. What is the greatest positive one week change?

Ⓐ $\$.20$ Ⓑ $\$.12$ Ⓒ $\$.04$

Ⓓ $\$.06$ Ⓔ $\$.10$

10. *Multiple Choice*   There was 26.2 inches of snow on the ground on Monday. On Tuesday 3 more inches fell. On Wednesday 4.2 inches melted off. On Thursday 2.6 inches melted. Friday 1.2 inches accumulated. How deep was the snow by Friday night?

Ⓐ 26.8 in   Ⓑ 19.2 in.   Ⓒ 23.6 in

Ⓓ 21.2 in.   Ⓔ 15.2 in.

11. *Multi-Step Problem*   Your science class is doing a fall bird count. Each day for a week the birds sighted by class members are counted. The results are as follows.

Mon	Tues.	Wed.	Thurs.	Fri.
52	50	61	59	48

a. Find the change in the number of birds counted from each day to the next.

b. What does a negative value represent?

c. What does a positive value represent?

Chapter 2

Standardized Test Practice

For use with pages 86–91

TEST TAKING STRATEGY **Spend no more than a few minutes on each question.**

1. **Multiple Choice** Each number in a matrix is called a(n) __?__ .

   Ⓐ matrices Ⓑ data Ⓒ entry

   Ⓓ variable Ⓔ constant

2. **Multiple Choice** Find the sum.

$$\begin{bmatrix} 5 & 6 \\ -2 & 1 \\ -6 & -3 \end{bmatrix} + \begin{bmatrix} -2 & 3 \\ -4 & 5 \\ 7 & 8 \end{bmatrix} = ?$$

   Ⓐ $\begin{bmatrix} 7 & 9 \\ 6 & 6 \\ 1 & 11 \end{bmatrix}$ Ⓑ $\begin{bmatrix} 3 & 9 \\ -6 & 6 \\ 1 & 5 \end{bmatrix}$

   Ⓒ $\begin{bmatrix} -3 & 9 \\ 8 & 5 \\ 1 & 5 \end{bmatrix}$ Ⓓ $\begin{bmatrix} -10 & 18 \\ 8 & 5 \\ -42 & -24 \end{bmatrix}$

   Ⓔ None of these

3. **Multiple Choice** Find the sum.

$$\begin{bmatrix} 5 & 8 & -6 \\ -7 & 3 & -2 \\ 14 & 2 & 1 \end{bmatrix} + \begin{bmatrix} 3 & 11 & -8 \\ 5 & -4 & -3 \\ 0 & 1 & -1 \end{bmatrix} = ?$$

   Ⓐ $\begin{bmatrix} 8 & 19 & -2 \\ -2 & -1 & -6 \\ 14 & 3 & 0 \end{bmatrix}$ Ⓑ $\begin{bmatrix} 8 & 19 & -2 \\ -2 & -1 & -5 \\ 14 & 3 & -2 \end{bmatrix}$

   Ⓒ $\begin{bmatrix} 8 & 19 & -14 \\ -2 & -1 & -5 \\ 14 & 3 & 0 \end{bmatrix}$ Ⓓ $\begin{bmatrix} 8 & 19 & -14 \\ -12 & -1 & 6 \\ 14 & 3 & 0 \end{bmatrix}$

   Ⓔ None of these

4. **Multiple Choice** Find the difference.

$$\begin{bmatrix} 6 & -8 & 7 & -5 \\ 1 & 9 & 0 & -6 \end{bmatrix} - \begin{bmatrix} 5 & 2 & -1 & 3 \\ 3 & -4 & 5 & -1 \end{bmatrix} = ?$$

   Ⓐ $\begin{bmatrix} 1 & -10 & 8 & -8 \\ -2 & 13 & -5 & -5 \end{bmatrix}$

   Ⓑ $\begin{bmatrix} 1 & -6 & 6 & -2 \\ -2 & 13 & 5 & -5 \end{bmatrix}$

   Ⓒ $\begin{bmatrix} 1 & -10 & 6 & -2 \\ -2 & 5 & 5 & -7 \end{bmatrix}$

   Ⓓ $\begin{bmatrix} 1 & -10 & 8 & -8 \\ -2 & 5 & -5 & 7 \end{bmatrix}$

   Ⓔ None of these

Multiple Choice Use the matrix below to answer Exercises 5 and 6.

Profit ($)

	Store 1	Store 2
January	15,283	10,272
February	9,865	5,950
March	10,525	-1,250

5. What was the amount of profit or loss of Store 2 in March?

   Ⓐ $5950 gain Ⓑ $5950 loss

   Ⓒ $1250 loss Ⓓ $1250 gain

   Ⓔ $10525 gain

6. What was the amount of profit or loss of Store 2 in February?

   Ⓐ $5950 gain Ⓑ $5950 loss

   Ⓒ $1250 loss Ⓓ $1250 gain

   Ⓔ $10525 gain

Quantitative Comparison In Exercises 7–9, use matrix A and matrix B to choose the statement that is the true about the given quantities.

   Ⓐ The quantity in column A is greater.

   Ⓑ The quantity in column B is greater.

   Ⓒ The two quantities are equal.

   Ⓓ The relationship cannot be determined from the information given.

   Matrix A
$$\begin{bmatrix} 2 & 3 & 6 & 8 \\ -1 & 7 & 3 & -4 \\ 5 & -8 & -12 & 0 \end{bmatrix}$$

   Matrix B
$$\begin{bmatrix} -7 & -3 & -2 & -5 \\ -2 & 8 & 7 & -9 \\ -6 & 9 & 11 & 6 \end{bmatrix}$$

	Column A	Column B
7.	The number of rows in Matrix A	The number of columns in matrix B
8.	The entry in the first row and first column of matrix A.	The entry in the first row and first column of matrix B
9.	The entry in the second row and first column of matrix A.	The entry in the second row and first column of matrix B.

NAME _____ DATE _____

Standardized Test Practice

For use with pages 93–98

TEST TAKING STRATEGY **Spend no more than a few minutes on each question.**

1. *Multiple Choice* Find the product of $(-8)(2)(-2)$.

   (A) 32   (B) -32 (C) -8

   (D) 8   (E) 18

2. *Multiple Choice* Find the product of $\left(-\frac{1}{2}\right)(-6)\left(-\frac{1}{3}\right)$.

   (A) $6\frac{1}{6}$ (B) $-6\frac{1}{6}$ (C) 1

   (D) -1 (E) $-\frac{6}{5}$

3. *Multiple Choice* Find the product of $(-3)(-x)(4)$.

   (A) $-12x$ (B) $12x$ (C) x

   (D) $-1x$ (E) $-12 - x$

4. *Multiple Choice* Find the product of $-(-x)(6)\left(\frac{1}{2}x\right)$.

   (A) $3x^2$ (B) $-3x^2$

   (C) $3 + 2x$ (D) $-3 + 2x$

   (E) $\frac{7}{2}x^2$

5. *Multiple Choice* The equation $a \cdot b = b \cdot a$ represents which property of multiplication?

   (A) commutative property

   (B) associative property

   (C) identity property

   (D) property of zero

   (E) property of opposites

6. *Multiple Choice* Evaluate the expression $(2x)(-3)(-1)$ for $x = -4$.

   (A) -24 (B) 24   (C) 18

   (D) -18 (E) -6

7. *Multiple Choice* Evaluate the expression $(3x)^2 - 8x$ when $x = -3$.

   (A) 12   (B) -57 (C) 105

   (D) 60   (E) 57

8. *Multiple Choice* A hawk dives towards the ground to catch a rabbit. It descends at a rate of -10.2 feet per second. What would be the vertical distance traveled in 5 seconds?

   (A) -15.2 ft   (B) 2.04 ft

   (C) -2.04 ft   (D) 51 ft

   (E) -51 ft

9. *Multiple Choice* A store is marking down its winter clothing for an end of season sale. It paid $50 for a fleece jacket and is now selling it for $43. The store has 30 jackets left to sell. If they sell all the jackets during the sale, how much will they lose?

   (A) $7   (B) $37   (C) $210

   (D) $1290   (E) $1500

10. *Multi-Step Problem* Your sister loans you $75 to buy a video game. You agree to work off your loan by doing her chores for $3 an hour.

    (A) Write an equation that you can use to determine the amount of money you still owe your sister.

    (B) How much do you owe after 12 hours of chores

    (C) How many hours of chores must you complete to pay for the $75 loan?

NAME _____ DATE _____

Standardized Test Practice

For use with pages 100–107

TEST TAKING STRATEGY   **Skip questions that are too difficult for you.**

1. *Multiple Choice*   Use the distributive property to rewrite $6(x - 3)$ without parentheses.

   Ⓐ $6x + 3$ ⠀⠀⠀ Ⓑ $6x - 18$

   Ⓒ $6x + 18$ ⠀⠀ Ⓓ $6x - 3$

   Ⓔ $x - 18$

2. *Multiple Choice*   Use the distributive property to rewrite $(3y - 9)(8y)$ without parentheses.

   Ⓐ $24y^2 - 72y$ ⠀ Ⓑ $24y^2 + 72y$

   Ⓒ $11y^2 - y$ ⠀⠀ Ⓓ $3y - 72y$

   Ⓔ $24y^2 - 9$

3. *Multiple Choice*   Choose the expression that is equivalent to $8x^2 - 4xy$.

   Ⓐ $(8x^2 - 4x)y$ ⠀ Ⓑ $(2x + y)(-4x)$

   Ⓒ $8\left(x^2 - \frac{1}{2}y\right)$ ⠀ Ⓓ $2x(6x - 2y)$

   Ⓔ $(2x - y)(4x)$

4. *Multiple Choice*   Simplify the expression $10x - 3y + 4y - 2x$.

   Ⓐ $8x - y$ ⠀⠀⠀ Ⓑ $12x + y$

   Ⓒ $12x + 7y$ ⠀⠀ Ⓓ $8x + y$

   Ⓔ $8x - 7y$

5. *Multiple Choice*   Simplify the expression $2x^2 - 5 + 3x^2 - 4x + 6$.

   Ⓐ $x + 1$ ⠀⠀⠀⠀⠀ Ⓑ $6x^2 - 4x + 1$

   Ⓒ $5x^2 - 4x + 11$ Ⓓ $5x^2 - 4x + 1$

   Ⓔ $2x^2 + 1$

6. *Multiple Choice*   Apply the distributive property, then simplify: $(4x - 2)(-5) + 3x$.

   Ⓐ $-17x - 10$ ⠀ Ⓑ $-23x - 10$

   Ⓒ $-17x + 10$ ⠀ Ⓓ $-23x + 10$

   Ⓔ $2x - 7$

7. *Multiple Choice*   Apply the distributive property, then simplify:
   $6y^2 - y(3y - 8) + 2y$.

   Ⓐ $3y^2 - 6y$ ⠀⠀ Ⓑ $3y^2 + 10y$

   Ⓒ $9y^2 - 6y$ ⠀⠀ Ⓓ $9y + 10y$

   Ⓔ $3y^2 + 2y - 8$

8. *Multiple Choice*   Write and simplify an expression modeling the area of the rectangle.

   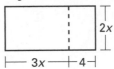

   Ⓐ $5x + 4$ ⠀⠀⠀⠀ Ⓑ $6x^2 + 8x$

   Ⓒ $5x^2 + 8x$ ⠀⠀ Ⓓ $14x^2$

   Ⓔ $13x^2$

9. *Multi-Step Problem*   You and your friends are going on a 20 mile canoe trip. The creek is low so you will have to portage, or carry, your canoe for some of the trip. You can travel 6 miles per hour in your canoe, and 0.7 mile per hour when portaging.

   a. Let p represent the miles you portage your canoe. Which function can you use to find the total time spent on your canoe trip.

   Ⓐ $T = 0.7p + 6(20 - p)$

   Ⓑ $T = \dfrac{0.7}{p} + \dfrac{6}{20 - p}$

   Ⓒ $T = \dfrac{p}{0.7} + \dfrac{20 - p}{6}$

   b. If you portage your canoe for 1.5 miles, how long does it take you to complete your trip?

NAME _____ DATE _____

Standardized Test Practice

For use with pages 108–113

TEST TAKING STRATEGY   Work as fast as you can through the easier problems, but not so fast that you make a careless mistakes.

1. *Multiple Choice* The expression $b \div a$ is equivalent to ___?___.

  Ⓐ $b \cdot \dfrac{a}{1}$ Ⓑ $b \cdot \dfrac{1}{a}$ Ⓒ $a \cdot \dfrac{b}{1}$

  Ⓓ $a \cdot \dfrac{1}{6}$ Ⓔ $\dfrac{a}{6}$

2. *Multiple Choice* Find the quotient.
$-39 \div 3\frac{1}{4} = ?$

  Ⓐ $-126\frac{3}{4}$ Ⓑ $126\frac{3}{4}$ Ⓒ -12

  Ⓓ 12 Ⓔ $-\frac{4}{3}$

3. *Multiple Choice* Find the quotient.
$-\frac{2}{3} \div \left(-\frac{3}{4}\right) = ?$

  Ⓐ $\frac{8}{9}$ Ⓑ $-\frac{8}{9}$ Ⓒ $\frac{1}{2}$

  Ⓓ $-\frac{1}{2}$ Ⓔ $\frac{6}{12}$

4. *Multiple Choice* Simplify the expression
$\dfrac{24x - 8}{4}$.

  Ⓐ $24x - 2$ Ⓑ $6x - 2$

  Ⓒ $6x - 8$ Ⓓ $6x + 2$

  Ⓔ $96x - 32$

5. *Multiple Choice* Simplify the expression
$\dfrac{-56x}{12y} \div \dfrac{8x^2}{6y}$.

  Ⓐ $\dfrac{-56x^3}{9y^2}$ Ⓑ $\dfrac{-2x}{7}$ Ⓒ $\dfrac{2x}{7}$

  Ⓓ $\dfrac{-7}{2x}$ Ⓔ $\dfrac{7}{2x}$

6. *Multiple Choice* Simplify the expression
$\dfrac{14x^2 - 21x}{7x}$.

  Ⓐ $2x^2 - 3x$ Ⓑ $7x - 14$

  Ⓒ $2x + 3$ Ⓓ $2x - 3$

  Ⓔ $7x^2 - 14$

7. *Multiple Choice* Simplify the expression
$-64x^3 \div \dfrac{16x^2}{5}$.

  Ⓐ $-20x$ Ⓑ $20x$ Ⓒ $-9x$

  Ⓓ $-204.8x^5$ Ⓔ $9x$

8. *Multiple Choice* Choose the value of x which cannot be contained in the domain of
$y = \dfrac{8}{5 - x}$.

  Ⓐ 0 Ⓑ 5 Ⓒ -5

  Ⓓ 3 Ⓔ -3

9. *Multiple Choice* A roller coaster descends from its highest point of 200 feet in 5 seconds. What is the velocity?

  Ⓐ -40 ft/sec   Ⓑ 40 ft/sec

  Ⓒ 0.025 ft/sec   Ⓓ -0.025 ft/sec

  Ⓔ -1000 ft/sec

Quantitative Comparison In Exercises 10–13, choose the statement that is true about the given quantities when $a = -2$ and $b = 3$.

  Ⓐ The quantity in column A is greater.

  Ⓑ The quantity in column B is greater.

  Ⓒ The two quantities are equal.

  Ⓓ The relationship can not be determined from the information given.

	Column A	Column B
10.	$\dfrac{a^2 - 3}{4}$	$\dfrac{18 - b^2}{b}$
11.	$\dfrac{-ab - 8}{a}$	$\dfrac{a}{b} \div \dfrac{b}{a}$
12.	$\dfrac{b - a^2}{3}$	$\dfrac{ab^2}{b - a}$
13.	$\dfrac{3a - 5b}{5}$	$\dfrac{5a^2 - b}{a}$

Algebra 1
Standardized Test Practice Workbook

LESSON

2.8

NAME _____ DATE _____

Standardized Test Practice

For use with pages 114–120

TEST TAKING STRATEGY **Spend no more than a few minutes on each question.**

1. *Multiple Choice* If an event has a probability of 0.75, what is the likelyhood of it occuring due to chance?
   - (A) Impossible
   - (B) Unlikely
   - (C) Quite likely
   - (D) Certain
   - (E) Likely to occur half the time.

2. *Multiple Choice* A car dealership has 12 cars, 6 trucks, and 7 sport utility vehicles on its lot. What is the probability of randomly choosing a truck?
   - (A) $\frac{12}{25}$
   - (B) $\frac{6}{25}$
   - (C) $\frac{7}{25}$
   - (D) $\frac{6}{19}$
   - (E) $\frac{7}{19}$

3. *Multiple Choice* You randomly choose an integer between 1 and 20. What are the odds that the integer is in the teens?
   - (A) $\frac{1}{20}$
   - (B) $\frac{1}{13}$
   - (C) $\frac{7}{13}$
   - (D) $\frac{7}{20}$
   - (E) $\frac{13}{20}$

4. *Multiple Choice* There are 20 dogs, 14 cats, 6 birds, 1 snake, and 2 rabbits at a local shelter. What are the odds of randomly choosing a cat?
   - (A) $\frac{7}{10}$
   - (B) $\frac{7}{17}$
   - (C) $\frac{14}{43}$
   - (D) $\frac{14}{29}$
   - (E) $\frac{1}{29}$

5. *Multiple Choice* Given that the probability of choosing a King from a deck of cards is $\frac{1}{13}$, find the odds.
   - (A) $\frac{1}{12}$
   - (B) $\frac{1}{13}$
   - (C) $\frac{4}{52}$
   - (D) $\frac{1}{14}$
   - (E) $\frac{4}{13}$

6. *Multiple Choice* You toss a coin 16 times. Six of the tosses were tails. What is the experimental probability of tossing a head?
   - (A) $\frac{3}{8}$
   - (B) $\frac{3}{5}$
   - (C) $\frac{5}{8}$
   - (D) $\frac{5}{3}$
   - (E) $\frac{1}{2}$

7. *Multiple Choice* A music store did a survey of 100 people to find out what type of music they bought. The survey showed 13 people prefered alternative, 23 blues, 25 hard rock, and 39 soft rock. If they surveyed one more person, what is the experimental probability that the person would prefer alternative?
   - (A) $\frac{1}{87}$
   - (B) $\frac{1}{4}$
   - (C) $\frac{13}{87}$
   - (D) $\frac{14}{100}$
   - (E) $\frac{13}{100}$

Quantitative Comparison In Exercises 8–10, refer to the table and choose the statement that is true about the given number.
   - (A) The quantity in column A is greater.
   - (B) The quantity in column B is greater.
   - (C) The two quantities are equal.
   - (D) The relationship can not be determined from the information given.

Students taking Algebra I

Grade	Female	Male
9th	26	22
10th	21	19
11th	16	23
12th	6	10

	Column A	Column B
8.	The probability of choosing a freshman male student	The probability of choosing a junior or senior female student
9.	The odds of choosing a junior.	The odds of choosing a sophmore.
10.	The probability of choosing a female.	The odds of choosing a female.

Chapter 2

Copyright © McDougal Littell Inc.
All rights reserved.

Algebra 1
Standardized Test Practice Workbook

15

NAME _____ DATE _____

Standardized Test Practice

For use with pages 132–137

TEST TAKING STRATEGY   Go back and check as much of your work as you can.

1. *Multiple Choice*   Choose the inverse operation of subtracting -17.
   Ⓐ Multiply by 17   Ⓑ Add -17
   Ⓒ Multiply by -17 Ⓓ Add 17
   Ⓔ Divide by -17

2. *Multiple Choice*   Choose the inverse operation of adding 8.
   Ⓐ Multiply by 8   Ⓑ Subtract 8
   Ⓒ Divide by 8   Ⓓ Subtract -8
   Ⓔ Multiply by -8

3. *Multiple Choice*   Solve $x - 6 = 10$.
   Ⓐ 4   Ⓑ -4 Ⓒ 16
   Ⓓ -16 Ⓔ 60

4. *Multiple Choice*   Solve $14 = y + 2$.
   Ⓐ 16   Ⓑ -16 Ⓒ -12
   Ⓓ 12   Ⓔ 7

5. *Multiple Choice*   Solve $8 + x = -10$.
   Ⓐ -18 Ⓑ 18   Ⓒ -2
   Ⓓ 2   Ⓔ 1.2

6. *Multiple Choice*   Solve $3 = |-10| - x$.
   Ⓐ 13   Ⓑ -13
   Ⓒ -7 Ⓓ -3
   Ⓔ 7

7. *Multiple Choice*   There are 22 students in a Spanish class. This is an increase of 8 students from last year. Which equation could be used to determine the number of students in last year's Spanish class?
   Ⓐ $22 + x = 8$ Ⓑ $x - 8 = 22$
   Ⓒ $22 + x = -8$ Ⓓ $x + 8 = 22$
   Ⓔ $22 + 8 = x$

8. *Multiple Choice*   There was 2.6 inches of rainfall today. The total precipitation for the month is now 8.2 inches. Which equation could be used to determine the amount of precipitation before today?
   Ⓐ $x - 2.6 = 8.2$ Ⓑ $2.6 + 8.2 = x$
   Ⓒ $x + 2.6 = 8.2$ Ⓓ $8.2 + x = 2.6$
   Ⓔ $2.6 - x = 8.2$

9. *Multiple Choice*   Find the length of the side x of the triangle. The perimeter is 14.
   Ⓐ 3
   Ⓑ 4
   Ⓒ 5
   Ⓓ 6
   Ⓔ 7

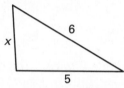

10. *Multiple Choice*   A pet store marks up the price of its puppies by $150. It sells them for $500 each. How much did the store originally pay for each puppy?
   Ⓐ $650   Ⓑ $350
   Ⓒ $250   Ⓓ $150
   Ⓔ $500

Quantitative Comparison   In Exercises 11–14, choose the statement below that is true about the given number.
   Ⓐ   The solution in column A is greater.
   Ⓑ   The solution in column B is greater.
   Ⓒ   The two solutions are equal.
   Ⓓ   The relationship cannot be determined from the given information.

	Column A	Column B
11.	$x + 2 = 6$	$y - 4 = -6$
12.	$x + |-3| = 3$	$6 = y + 4$
13.	$|x| - 2 = 6$	$y - 3 = 4$
14.	$-6 - x = 3$	$-2 = y - (-8)$

TEST TAKING STRATEGY   Before you give up on a question, try to eliminate some of your choices as you can make an educated guess.

1. *Multiple Choice*   Choose the inverse operation of multiplying by -6.
   - (A) Multiply by 6
   - (B) Divide by 6
   - (C) Divide by -6
   - (D) Add 6
   - (E) Add -6

2. *Multiple Choice*   Which one of these steps can by used to solve the equation $\frac{2}{3}x = 8$?
   - **I.** Multiply by $\frac{2}{3}$
   - **II.** Multiply by $\frac{3}{2}$
   - **III.** Divide by $\frac{2}{3}$
   - **IV.** Divide by $\frac{3}{2}$
   - (A) II only
   - (B) III only
   - (C) I and III
   - (D) II and III
   - (E) II and IV

3. *Multiple Choice*   Solve $18 = \dfrac{x}{-3}$.
   - (A) -6
   - (B) 6
   - (C) -54
   - (D) 54
   - (E) 21

4. *Multiple Choice*   Solve $-\frac{1}{3}y = \frac{4}{5}$.
   - (A) $-2\frac{2}{5}$
   - (B) $2\frac{2}{5}$
   - (C) $-\frac{5}{12}$
   - (D) $\frac{5}{12}$
   - (E) $1\frac{2}{15}$

5. *Multiple Choice*   If $-0.3x = 12$, then $x = \underline{\ ?\ }$.
   - (A) -40
   - (B) 40
   - (C) 3.6
   - (D) -3.6
   - (E) 12.3

6. *Multiple Choice*   You're installing a new fence around an inground swimming pool. The pool is a rectangle measuring 12 feet by 20 feet. The fence must be the same distance away from the pool on all sides. If you have 144 feet of fence, what is the distance from any side of the pool to the fence. (*Hint:* Draw a diagram.)
   - (A) 20 ft
   - (B) 10 ft
   - (C) 5 ft
   - (D) 40 ft
   - (E) 15 ft

7. *Multiple Choice*   The two triangles below are similar triangles. What is the length of $\overline{DE}$?

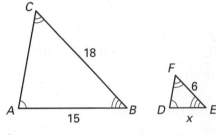

   - (A) 3
   - (B) 4
   - (C) 15
   - (D) 10
   - (E) 5

8. *Multiple Choice*   If $a = b$ and $c \neq 0$, then $\dfrac{a}{c} = \dfrac{b}{c}$ represents which property of equality?
   - (A) Addition property of equality
   - (B) Subtraction property of equality
   - (C) Multiplication property of equality
   - (D) Division property of equality
   - (E) Reciprical property of equality

Quantitative Comparison   In Exercises 9–11, choose the statement that is the true about the given number.
   - (A) The soution in column A is greater.
   - (B) The solution in column B is greater.
   - (C) The two solutions are equal.
   - (D) The relationship cannot be determined from the information given.

	Column A	Column B
9.	$-8x = 24$	$\frac{1}{2}y = -4$
10.	$-3 = -\frac{1}{4}x$	$\frac{2}{3}y = 8$
11.	$\frac{1}{5}x = 3\frac{1}{4}$	$-\frac{5}{6} = -\frac{1}{3}y$

Chapter 3

NAME _____ DATE _____

Standardized Test Practice

For use with pages 145–152

TEST TAKING STRATEGY **Think positively during a test. This will help keep up your confidence and enable you to focus on each question.**

1. *Multiple Choice* Which one of these steps should you use first to solve $5x - 2 = 15$?

   I. Subtract 2   II. Add 2

   III. Divide by 5   IV. Multiply by 5

   Ⓐ I   Ⓑ II

   Ⓒ III   Ⓓ II or III

   Ⓔ I or III

2. *Multiple Choice* Solve $-7 = \frac{2}{3}x + 5$.

   Ⓐ -8 Ⓑ -3 Ⓒ -18

   Ⓓ 18   Ⓔ $-\frac{4}{3}$

3. *Multiple Choice* Solve $6x - 5 - 3x = 4$.

   Ⓐ 3   Ⓑ -3 Ⓒ 1

   Ⓓ -1 Ⓔ $\frac{1}{3}$

4. *Multiple Choice* Solve $-4x - 6 - 3x = 5$.

   Ⓐ $-\frac{11}{7}$ Ⓑ -3 Ⓒ $1\frac{6}{11}$

   Ⓓ 3   Ⓔ $2\frac{2}{11}$

5. *Multiple Choice* Solve the equation $8(2x - 1) - 5x = 25$.

   Ⓐ -11 Ⓑ -3 Ⓒ $1\frac{6}{11}$

   Ⓓ 3   Ⓔ $2\frac{2}{11}$

6. *Multiple Choice* Solve $-\frac{2}{3}(x + 2) = 12$.

   Ⓐ -10 Ⓑ -6 Ⓒ -16

   Ⓓ $10\frac{2}{3}$ Ⓔ -20

7. *Multiple Choice* Solve the equation $32x - 4(7x + 3) = 16$.

   Ⓐ $\frac{17}{29}$ Ⓑ -7 Ⓒ 1

   Ⓓ 7   Ⓔ -1

8. *Multiple Choice* You live in the United States close to the Canadian border. A Canadian weatherman reports your temperature as being 25°C. Use the formula $F = \frac{9}{5}C + 32$ to find the temperature in degrees Fahrenheit.

   Ⓐ 46°F   Ⓑ 77°F   Ⓒ -4°F

   Ⓓ 102°F   Ⓔ 68°F

9. *Multiple Choice* You billed your neighbor $52 for work you did around his house. Materials cost you $16, and you charged $4 an hour. How many hours did you work?

   Ⓐ 7 hours   Ⓑ 8 hours

   Ⓒ 9 hours   Ⓓ 17 hours

   Ⓔ 32 hours

Quantitative Comparison In Exercises 10–13, choose the statement that is the true about the given number.

   Ⓐ The soution in column A is greater.

   Ⓑ The solution in column B is greater.

   Ⓒ The two solutions are equal.

   Ⓓ The relationship cannot be determined from the information given.

	Column A	Column B
10.	$2x - 7 = -15$	$-3y + 6 = 18$
11.	$3x - 2(x + 4) = 8$	$\frac{2}{5}(y - 4) = 6$
12.	$14(2 - x) = 35$	$12y + 7 - 5y = -14$
13.	$y = 2x + 3$	$x = 2y + 3$

Chapter 3

NAME _____ DATE _____

Standardized Test Practice

For use with pages 154–159

TEST TAKING STRATEGY   **Be aware of how much time you have left, but keep focused on your work.**

1. *Multiple Choice*   Solve $3x - 6 = 4x + 8$.

    Ⓐ -2 **Ⓑ** 2 **Ⓒ** -14

    Ⓓ 14 **Ⓔ** -1

2. *Multiple Choice*   Solve the equation $5x - 12 = -3x + 6$.

    Ⓐ $-2\frac{1}{4}$ **Ⓑ** $2\frac{1}{4}$ **Ⓒ** $\frac{3}{4}$

    Ⓓ $-\frac{3}{4}$ **Ⓔ** 9

3. *Multiple Choice*   Solve the equation $\frac{1}{4}(16x - 8) + 6 = 12 - 5x$.

    Ⓐ $\frac{8}{9}$ **Ⓑ** $1\frac{1}{8}$ **Ⓒ** $-\frac{8}{9}$

    Ⓓ $1\frac{1}{9}$ **Ⓔ** 14

4. *Multiple Choice*   Solve $x + 3 = x - 6$.

    Ⓐ 3 **Ⓑ** 6 **Ⓒ** -3

    Ⓓ -6 **Ⓔ** No solution

5. *Multiple Choice*   Solve the equation $4x - 3(x + 2) = 5(4 - x)$.

    Ⓐ $-4\frac{1}{3}$ **Ⓑ** $4\frac{1}{3}$ **Ⓒ** $\frac{3}{13}$

    Ⓓ $-\frac{3}{13}$ **Ⓔ** No solution

6. *Multiple Choice*   Solve the equation $-10 + 2x + 3(5 - x) = \frac{1}{2}(4x - 8)$.

    Ⓐ -3 **Ⓑ** 3 **Ⓒ** $\frac{1}{3}$

    Ⓓ $-\frac{1}{3}$ **Ⓔ** 13

7. *Multiple Choice*   A local health club has a large swimming pool. Non-members must pay a $6 per day fee to use the pool. Members of the club pay $1, but must pay a yearly membership fee of $250. How many days must you use the pool to justify joining the club?

    Ⓐ 36 days   **Ⓑ** 38 days

    Ⓒ 50 days   **Ⓓ** 40 days

    Ⓔ 45 days

8. *Multiple Choice*   Your club decides to sell boxes of stationery to raise money. Each box costs you $2.50 and there is a one time delivery fee of $30. You plan to sell the stationery for $4.50 per box. Which equation should you use to determine how many boxes you must sell to cover your costs?

    Ⓐ $4.50x + 30 = 2.50x$

    Ⓑ $2.50x - 30 = 4.50x$

    Ⓒ $30 - 2.50x = 4.50x$

    Ⓓ $2.50x + 30 = 4.50x$

    Ⓔ $30 - 4.50x = 2.50x$

Quantitative Comparison   In Exercises 9–12, choose the statement that is the true about the given number.

    Ⓐ The solution in column A is greater.

    Ⓑ The solution in column B is greater.

    Ⓒ The two solutions are equal.

    Ⓓ The relationship cannot be determined from the information given.

	Column A	Column B
9.	$4x + 8 = 2(2x + 4)$	$3x = 7x - 6$
10.	$16 - 2x = 2x$	$\frac{1}{2}(x - 4) = 4x$
11.	$6x - 1 = 5x + 1$	$4x - 6 = 7x - 12$
12.	$\frac{1}{3}(3x - 6) = 2x$	$3x - 7 - 5x = 5 + x$

NAME _____ DATE _____

Standardized Test Practice

For use with pages 160–165

TEST TAKING STRATEGY **Some questions involve more than one step. Reading too quickly might lead to mistaking the answer to a preliminary step for your final answer.**

Multiple Choice For Exercises 1–5, use the following information. You are planting a garden with 3 varieties of flowers. Your garden space is 36 inches deep by 72 inches wide. You want to leave 4 inches of space all around the outside, and 3 inches between flower varieties. Assume each flower bed will be the same width.

1. Which equation can be used to find the width of each variety of flower bed?

- Ⓐ $3x + 14 = 72$
- Ⓑ $3x + 14 = 36$
- Ⓒ $3x + 17 = 72$
- Ⓓ $3x + 10 = 72$
- Ⓔ $3x + 17 = 36$

2. To the nearest tenth of an inch, how wide can each variety of flower be?

- Ⓐ 7.3 in.
- Ⓑ 19.3 in.
- Ⓒ 20.7 in.
- Ⓓ 18.3 in.
- Ⓔ 6.3 in.

3. Which equation can be used to find the depth of each flower bed?

- Ⓐ $8 + 3x = 36$
- Ⓑ $4 + x = 36$
- Ⓒ $8 + x = 36$
- Ⓓ $8 + x = 72$
- Ⓔ $8 + 3x = 72$

4. How deep can each type of flower planting be?

- Ⓐ 9.3 in.
- Ⓑ 32 in.
- Ⓒ 21.3 in.
- Ⓓ 64 in.
- Ⓔ 28 in.

5. If you double the width of your flower bed, how wide (to the nearest tenth of an inch) can each variety of flower bed be?

- Ⓐ 38.6 in.
- Ⓑ 36.6 in.
- Ⓒ 14.6 in.
- Ⓓ 43.3 in.
- Ⓔ 56 in.

6. *Multiple Choice* You have $210 and save $5 a week. Your friend has $265, saves nothing, and spends $10 a week. Determine how many weeks it will take before you have more money that your friend. Which one of the choices below can be used to check your answer?

I.

II.

Week	0	1	2	3	4
You	210	215	220	225	230
Friend	265	255	245	235	225

III.

Week	1	2	3	4	5
You	+5	+10	+15	+20	+25
Friend	−10	−20	−30	−40	−50

- Ⓐ I only
- Ⓑ II only
- Ⓒ III only
- Ⓓ I and II
- Ⓔ I and III

7. *Multi-Step Problem* Two cross-country skiers leave the ski lodge at the same time, and travel the same trail. The first skier returns to the lodge after $2\frac{1}{2}$ hours and averaged 8 miles per hour. The second skier returns 195 minutes later.

a. Draw a diagram modeling the situation.

b. Find the second skier's speed.

c. Find the distance traveled by the skiers

d. The second skier had to stop for 15 minutes to wax his skis. What was the speed of this skier if you consider only skiing time?

Chapter 3

NAME _____ DATE _____

Standardized Test Practice

For use with pages 166–172

TEST TAKING STRATEGY   **Read all of the choices before deciding which is the correct one.**

1. *Multiple Choice*   Evaluate 4.36(7.32). Round to the nearest tenth.
   - **A** 31.91
   - **B** 31.92
   - **C** 31.9
   - **D** 32.0
   - **E** 31.0

2. *Multiple Choice*   Solve $17x - 19 = 106$. Round to the nearest hundredth.
   - **A** 5.12
   - **B** 5.118
   - **C** 7.36
   - **D** 5.11
   - **E** 7.35

3. *Multiple Choice*   Solve $-36 - 19x = 14$. Round to the nearest tenth.
   - **A** 2.6
   - **B** 2.7
   - **C** 1.16
   - **D** 1.2
   - **E** −2.6

4. *Multiple Choice*   Solve the equation $14x - 8 = -33 + 2x$. Round to the nearest hundredth.
   - **A** 2.08
   - **B** −3.42
   - **C** −1.56
   - **D** −2.08
   - **E** −2.09

5. *Multiple Choice*   Solve the equation $-3(4.2 - 3.5x) = 8x - 7.2$. Round to the nearest thousandth.
   - **A** 1.070
   - **B** 2.160
   - **C** −0.292
   - **D** −0.291
   - **E** 7.920

6. *Multiple Choice*   Solve the equation $0.2(-3.1x - 2) = -1.1(2x - 3)$. Round to the nearest tenth.
   - **A** 2.3
   - **B** −2.3
   - **C** 1.84
   - **D** 1.8
   - **E** 2.34

7. *Multiple Choice*   Solve the equation $-1.6(2.3x) = 5.7(-3.6x - 0.3)$. Round to the nearest tenth.
   - **A** −9.8
   - **B** −0.1
   - **C** −14.15
   - **D** 0.1
   - **E** −0.2

8. *Multiple Choice*   You are shopping for a birthday present. You have $20.00 to spend and your state's sales tax is 6%. Which algebraic model could be used to determine your price limit for the present?
   - **A** $0.06x = 20$
   - **B** $1.06x = 20$
   - **C** $x = 20(0.06)$
   - **D** $x = 20(1.06)$
   - **E** none of these

9. *Multiple Choice*   Determine the maximum amount of money you can spend for the present in Exercise 9.
   - **A** $12.00
   - **B** $18.87
   - **C** $18.86
   - **D** $18.80
   - **E** $21.20

10. *Multi-Step Problem*   Your cellular phone company charges $29.95 per month and gives you 30 minutes of free calls per month. Additional minutes cost $.32 per minute. Your bill last month came to $46.82. Let x represent the number of minutes you talked last month.

    a. Write an equation that models the situation.

    b. Approximately how many minutes did you use the phone? Round to the nearest tenth.

    c. How many minutes did you have to pay extra money for?

Standardized Test Practice

For use with pages 174–179

TEST TAKING STRATEGY   Learn as much as you can about a test ahead of time, such as the types of questions and the topics that the test will cover.

1. **Multiple Choice**   Using the formula for the area of a rectangular prism, $V = lwh$, solve for w.

   (A) $w = Vlh$ (B) $w = V - l - h$

   (C) $w = \dfrac{lh}{V}$ (D) $w = V - lh$

   (E) $w = \dfrac{V}{lh}$

2. **Multiple Choice**   Using the formula for interest, $I = Prt$, solve for r.

   (A) $r = \dfrac{Pt}{I}$ (B) $r = \dfrac{I}{Pt}$

   (C) $r = I - Pt$ (D) $r = I - P - t$

   (E) $r = IPt$

3. **Multiple Choice**   Using the formula for the volume of a circular cylinder, $V = \pi r^2 h$, solve for h.

   (A) $h = \dfrac{V}{\pi r^2}$ (B) $h = \dfrac{\pi r^2}{V}$

   (C) $h = V\pi r^2$ (D) $h = V - \pi - r^2$

   (E) $h = V - \pi r^2$

4. **Multiple Choice**   Find the height of a triangle if the base measures 6 inches and the area is 15 square inches. Use the formula for the area of a triangle, $A = \frac{1}{2}bh$.

   (A) 3 in.   (B) 4 in.

   (C) 5 in.   (D) 6 in.

   (E) 7 in.

5. **Multiple Choice**   Rewrite the equation $3x + 2y = 10$ so that y is a function of x.

   (A) $y = -3x + 5$ (B) $x = -\dfrac{2y}{3} + \dfrac{10}{3}$

   (C) $y = -\dfrac{3}{2}x + 5$ (D) $x = -2y + 7$

   (E) $2y = 10 - 3x$

6. **Multiple Choice**   How long will it take $1000 invested in a savings account to double if the interest rate is 8%? Use the formula $I = Prt$.

   (A) 5 yrs   (B) 11.5 yrs

   (C) 6.25 yrs   (D) 10 yrs

   (E) 12.5 yrs

7. **Multiple Choice**   Rewrite the equation $3y - 2(x + 1) = y + 3(x + 6)$ so that y is a function of x.

   (A) $y = 2x + 10$ (B) $y = \frac{1}{2}x + 10$

   (C) $x = \frac{2}{5}y - 4$ (D) $y = \frac{5}{2}x + 10$

   (E) $x = \frac{4}{5}y - 4$

Quantitative Comparison   In Exercises 8–11, choose the statement that is true about the given number.

   (A)   The quantity in column A is greater.

   (B)   The quantity in column B is greater.

   (C)   The two quantities are equal.

   (D)   The relationship can not be determined from the information given.

	Column A	Column B
8.	The value of r in the formula $d = rt$ when $d = 25$ and $t = 6$.	The value of r in the formula $d = rt$ when $d = 36$ and $t = 8$.
9.	The value of r in the formula $A = \pi r^2$ when $A = 78.5$.	The value of r in the formula $C = 2\pi r$ when $C = 157$.
10.	The value of y in the function $y = 2x - 10$ when $x = 5$.	The value of y in the function $y = \frac{1}{2}x - 8$ when $x = 16$.
11.	The value of y in the function $y = \frac{2}{3}x + 12$ when $x = 9$.	The value of y in the equation $2y - x = 3y - 12$ when $x = 2$.

Chapter 3

Standardized Test Practice

For use with pages 180–186

TEST TAKING STRATEGY   Before you give up on a question, try to eliminate some of your choices so you can make an educated guess.

1. *Multiple Choice*   Which would be the correct unit rate for $3 for 12 cards?

(A) $3 per card   (B) 3 cards per dollar

(C) $.25 per dozen (D) $.25 per card

(E) None of these

2. *Multiple Choice*   Which model would you use to change 12 minutes to seconds?

(A) $12 \text{ min} \cdot \dfrac{1 \text{ min}}{60 \text{ sec}}$ (B) $12 \text{ min} \cdot \dfrac{60 \text{ sec}}{1 \text{ min}}$

(C) $12 \text{ min} \cdot \dfrac{60 \text{ min}}{1 \text{ sec}}$ (D) $12 \text{ min} \cdot \dfrac{1 \text{ sec}}{60 \text{ min}}$

(E) None of these

3. *Multiple Choice*   Find the average speed of a person boating 126 miles in 5 hours.

(A) 25.2 miles   (B) 25.2 mi/hr

(C) 630 mi/hr   (D) 3.9 mi/hr

(E) 3.9 hours per mile

4. *Multiple Choice*   Convert 250 pesos to dollars. The exchange rate is 9.99 pesos per United States dollar.

(A) $2497.50   (B) $259.99

(C) $25.03   (D) $240.01

(E) $30.02

5. *Multiple Choice*   Which model would you use to convert 35 miles per hour to feet per second?

(A) $35 \text{ mi/hr} \times \dfrac{5280 \text{ ft}}{1 \text{ mile}} \times \dfrac{60 \text{ min}}{1 \text{ hr}}$

(B) $35 \text{ mi/hr} \times \dfrac{1 \text{ mile}}{5280 \text{ ft}} \times \dfrac{1 \text{ hr}}{60 \text{ min}}$

(C) $35 \text{ mi/hr} \times \dfrac{5280 \text{ ft}}{1 \text{ mile}} \times \dfrac{1 \text{ hr}}{360 \text{ sec}}$

(D) $35 \text{ mi/hr} \times \dfrac{5280 \text{ ft}}{1 \text{ mile}} \times \dfrac{1 \text{ hr}}{3600 \text{ sec}}$

(E) $35 \text{ mi/hr} \times \dfrac{5280 \text{ ft}}{1 \text{ mile}} \times \dfrac{1 \text{ hr}}{60 \text{ min}}$

6. *Multiple Choice*   If 19% of $x = 45$, then $x = $ __?__. Round to the nearest tenth.

(A) 236.8   (B) 236.9

(C) 8.5   (D) 8.6

(E) 0.4

7. *Multiple Choice*   What percent of people were in favor of a survey question if 162 people voted no out of a total of 525 voters? Round to the nearest tenth.

(A) 69%   (B) 31%

(C) 30.8%   (D) 69.1%

(E) 30.9%

8. *Multiple Choice*   What is the total charge of a meal before tip if the tip came to $5.25 and was 15% of your bill?

(A) $78.75   (B) $3.64

(C) $40   (D) $79

(E) $35

9. *Multi-Step Problem*   Your a stocker at a grocery store. The store received a shipment of 500 heads of lettuce. You sample 50 heads of lettuce and found that 4 heads were brown.

a. Write a ratio that would estimate the number of heads of lettuce in the shipment that were brown.

b. Use the ratio from above to estimate the number of brown heads in the shipment.

c. If the store has a policy of accepting any shipment of produce if damaged produce if 7% or less of the shipment, should you reject or accept this shipment?

Chapter 3

Standardized Test Practice

For use with pages 203–208

TEST TAKING STRATEGY **Read all of the answer choices before deciding which is the correct one.**

Multiple Choice For Exercises 1–5, refer to the coordinate plane below

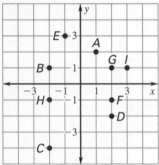

1. Which quadrant is point *B* in?
   - Ⓐ Quadrant I
   - Ⓑ Quadrant II
   - Ⓒ Quadrant III
   - Ⓓ Quadrant IV
   - Ⓔ Cannot be determine from the given information.

2. Which ordered pair corresponds to the point labeled *A*?
   - Ⓐ (1, 2)
   - Ⓑ (2, 1)
   - Ⓒ (−1, 2)
   - Ⓓ (−1, −2)
   - Ⓔ (2, −1)

3. Which point corresponds to the ordered pair (2, −1)?
   - Ⓐ *G*
   - Ⓑ *B*
   - Ⓒ *H*
   - Ⓓ *D*
   - Ⓔ *F*

4. Which quadrant would the ordered pair (−3, −2) be in?
   - Ⓐ I
   - Ⓑ II
   - Ⓒ III
   - Ⓓ IV
   - Ⓔ II or III

5. Which points have positive *y*-coordinates?
   - Ⓐ *A* and *F*
   - Ⓑ *A* and *B*
   - Ⓒ *B* and *H*
   - Ⓓ *B* and *F*
   - Ⓔ *H* and *F*

6. *Multiple Choice* In the scatter plot below, what conclusion can you come to from the data plotted?

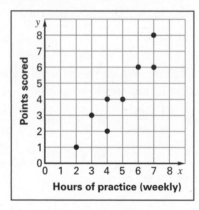

Hours of practice (weekly)

   - Ⓐ There is a strong positive correlation.
   - Ⓑ There is a weak positive correlation.
   - Ⓒ There is no correlation.
   - Ⓓ There is a weak negative correlation.
   - Ⓔ There is a strong negative correlation.

7. *Multi-Step Problem* The number of customers (in thousands) at a winter resort are shown in the table.

Year	1992	1993	1994
Skiers	8.2	8.6	9.1
Snowboarders	2.1	3	4.2

Year	1995	1996	1997
Skiers	9.7	10.3	10.8
Snowboarders	4.9	5.9	6.9

   a. Construct a scatter plot of the data

   b. Describe the pattern of the number of skiers

   c. Describe the pattern of the number of snowboarders.

   d. Predict the number of snowboarders for 2001.

NAME _____ DATE _____

Standardized Test Practice

For use with pages 210–217

TEST TAKING STRATEGY   Learn as much as you can about a test ahead of time, such as the types of questions and the topics that the test will cover.

1. *Multiple Choice*   Choose the ordered pair that is a solution of $2x - 3y = 6$.

    A $(5, 1)$ **B** $(0, 6)$ **C** $(4, 3)$

    D $(6, 2)$ **E** $(3, 2)$

2. *Multiple Choice*   Choose the ordered pair that is a solution of $y = -3$.

    A $(1, 3)$ **B** $(-1, -3)$

    C $(-3, 0)$ **D** $(3, -1)$

    E $(-3, -1)$

3. *Multiple Choice*   Choose the group of ordered pairs that are solutions of the equation $y = -3x + 4$.

    A $(1, 0), (2, -2)$ **B** $(1, 2), (3, 4)$

    C $(1, 1), (3, -5)$ **D** $(0, 4), (2, 2)$

    E $(2, -3), (4, 5)$

4. *Multiple Choice*   Choose the group of ordered pairs that are solutions of the equation $y = 8\left(\frac{1}{4}x - 2\right)$.

    A $(0, -16), (2, -12)$

    B $(0, -14), (1, -12)$

    C $(1, -16), (4, -8)$

    D $(4, -8), (8, 8)$

    E $(8, 8), (2, -12)$

5. *Multiple Choice*   Which point does *not* lie on the graph of $x = -2$?

    A $(-2, 0)$ **B** $(-2, -2)$

    C $(2, -2)$ **D** $(-2, 4)$

    E $(-2, 1)$

6. *Multiple Choice*   Which point does *not* lie on the graph of $3y + \frac{1}{2}x = 8$?

    A $\left(2, \frac{7}{3}\right)$ **B** $\left(0, \frac{8}{3}\right)$ **C** $\left(1, \frac{6}{3}\right)$

    D $(4, 2)$ **E** $(10, 1)$

7. *Multiple Choice*   What is the equation of the line shown?

    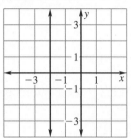

    A $x = -2$ **B** $x = 2$

    C $y = -2$ **D** $y = 2$

    E $y = 0$

8. *Multiple Choice*   What is the equation of the line shown?

    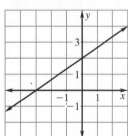

    A $2x + 3y = 6$ **B** $2x + 3y = -6$

    C $-2x + 3y = 6$ **D** $-2x - 3y = 6$

    E $2x - 3y = -6$

9. *Multi-Step Problem*   Your parents limit your phone use to 18 hours every week. You need to split that between hours spent on the phone with your friends and hours spent using the internet. Your parents count one hour on line as $\frac{1}{2}$ hour of phone time since some of your time is spent doing homework. An algebraic model is $x + \frac{1}{2}y = 18$ where x is hours spent on the phone and y is the hours spent on line.

    a. Solve the equation for y.

    b. Use the equation from part (a) to make a table of values for $x = 5$, $x = 10$, and $x = 15$.

    c. Plot the points and draw the line.

Chapter 4

NAME _____ DATE _____

Standardized Test Practice

For use with pages 218–224

TEST TAKING STRATEGY   Some questions involve more than one step. Reading too quickly might lead to mistaking the answer to a preliminary step for your final answer.

1. *Multiple Choice*   What is the *x*-intercept of the line $5x - 2y = 10$?

    Ⓐ 0   Ⓑ 2   Ⓒ 5
    Ⓓ −2   Ⓔ −5

2. *Multiple Choice*   What is the *y*-intercept of the line $5x - 2y = 10$?

    Ⓐ 0   Ⓑ 2   Ⓒ 5
    Ⓓ −2   Ⓔ −5

3. *Multiple Choice*   What is the *x*-intercept of the line $\frac{1}{2}x + \frac{3}{5}y = 20$?

    Ⓐ 0   Ⓑ 12   Ⓒ $\frac{100}{3}$
    Ⓓ 10   Ⓔ 40

4. *Multiple Choice*   What is the *y*-intercept of the line shown?

    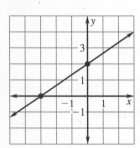

    Ⓐ 3   Ⓑ −3   Ⓒ 2
    Ⓓ −2   Ⓔ 0

5. *Multiple Choice*   What is the *x*-intercept of the line shown in exercise 4?

    Ⓐ 3   Ⓑ −3   Ⓒ 2
    Ⓓ −2   Ⓔ 0

6. *Multiple Choice*   A hotdog stand sells foot-long hotdogs fir $2.50 and 6 inch hotdogs for $1.50. If the stand makes $54 one day, choose the equation which models the situation.

    Ⓐ $2.50x - 1.50y = 54$
    Ⓑ $2.50x + 1.50y = 54$
    Ⓒ $y = -1.67x - 36$
    Ⓓ $y = -1.67x + 36$
    Ⓔ B and D

Quantitative Comparison   In Exercises 7–10, choose the statement that is true about the given number.

Ⓐ   The number in column A is greater

Ⓑ   The number in column B is greater

Ⓒ   The two numbers are equal

Ⓓ   The relationship cannot be determined from the information given.

	Column A	Column B
7.	The *x*-intercept of $3y + 6x = 12$	The *y*-intercept of $2y + 7x = 14$
8.	The *x*-intercept of $y = 6$	The *y*-intercept of $x = 3$.
9.	0	The *y*-intercept of $x + \frac{1}{2}y = 4$
10.	The *x*-intercept of $\frac{1}{2}x - 2y = 8$	The *y*-intercept of $\frac{1}{2}x - 2y = 8$

Chapter 4

NAME _____ DATE _____

Standardized Test Practice

For use with pages 226–233

TEST TAKING STRATEGY   Spend no more than a few minutes on each question.

1. *Multiple Choice* Which is the formula to find the slope of a line?

(A) $m = \dfrac{x_2 - x_1}{y_2 - y_1}$ **(B)** $m = \dfrac{y_2 + y_1}{x_2 + x_1}$

(C) $m = \dfrac{y_2 - y_1}{x_2 - x_1}$ **(D)** $m = \dfrac{x_2 + x_1}{y_2 - y_1}$

(E) $m = \dfrac{y_2 - x_2}{y_1 - x_1}$

2. *Multiple Choice* Find the slope of the line passing through the points $(5, 8)$ and $(9, 4)$.

(A) 3   **(B)** 1   **(C)** $\frac{3}{4}$

(D) -1 **(E)** $\frac{4}{3}$

3. *Multiple Choice* Find the slope of the line passing through the points $(-6, 7)$ and $(2, 9)$.

(A) $\frac{1}{4}$ **(B)** -2 **(C)** 4

(D) -4 **(E)** $-\frac{1}{2}$

4. *Multiple Choice* Find the slope of the line passing through the points $(3, 4)$ and $(-2, 4)$.

(A) undefined   **(B)** 0   **(C)** 8

(D) $\frac{8}{5}$ **(E)** $\frac{5}{8}$

5. *Multiple Choice* Find the slope of the line passing through the points $(7, -2)$ and $(7, 9)$.

(A) undefined   **(B)** 0   **(C)** $\frac{11}{14}$

(D) $\frac{1}{2}$ **(E)** 2

6. *Multiple Choice* Find the value of y so that the line passing through $(2, 6)$ and $(1, y)$ has a slope of 5.

(A) $\frac{5}{29}$ **(B)** 9   **(C)** $\frac{27}{5}$

(D) -1 **(E)** 1

7. *Multiple Choice* Find the value of x so that the line passing through $(10, 5)$ and $(x, 9)$ has a slope of -2.

(A) 2   **(B)** -2 **(C)** 8

(D) -8 **(E)** -17

8. *Multiple Choice* In 1989, a movie ticket cost \$3.00. In 1999, a movie tickets cost \$5.50. Find the average rate of change of movie tickets price in dollars per year.

(A) \$.20   **(B)** \$.25   **(C)** \$.30

(D) \$.85   **(E)** \$4.00

Quantitative Comparison In Exercises 9–12, choose the statement that is true about the given number.

(A) The number in column A is greater

(B) The number in column B is greater

(C) The two numbers are equal

(D) The relationship cannot be determined from the information given.

	Column A	Column B
9.	The slope of the line through $(6, 6)$ and $(10, 8)$	The slope of the line through $(5, 8)$ and $(17, 14)$
10.	The slope of the line through $(3, 6)$ and $(5, 3)$	0
11.	The slope of the line through $(1, 3)$ and $(8, 10)$	The slope of the line through $(12, 6)$ and $(18, 0)$
12.	The slope of the line through $(5, y)$ and $(7, 3)$	The slope of the line through $(4, 8)$ and $(x, 14)$

NAME _____ DATE _____

Standardized Test Practice

For use with pages 234–239

TEST TAKING STRATEGY   **Go back and check as much of your work as you can.**

1. *Multiple Choice*   In a direct variation model, if the constant of variation in the equation is 3, then the slope is ___?___.

Ⓐ 3   Ⓑ -3

Ⓒ $\frac{1}{3}$ Ⓓ $-\frac{1}{3}$

Ⓔ Cannot be determined from the given information.

2. *Multiple Choice*   Choose the constant of variation of the direct variation model $y = -5x$.

Ⓐ 5   Ⓑ -5 Ⓒ $-\frac{1}{5}$

Ⓓ x Ⓔ y

3. *Multiple Choice*   Choose the constant of variation of the direct variation model $-\frac{1}{2}x = y$.

Ⓐ x Ⓑ -2 Ⓒ $\frac{1}{2}$

Ⓓ $-\frac{1}{2}$ Ⓔ y

4. *Multiple Choice*   The variables x and y vary directly. Which equation relates x and y when $x = 4$ and $y = 12$?

Ⓐ $x = 4y$ Ⓑ $y - 4x$

Ⓒ $x = 3y$ Ⓓ $y - 12x$

Ⓔ $y = 3x$

5. *Multiple Choice*   The variables x and y vary directly. Which equation relates x and y when $x = \frac{1}{2}$ and $y = 6$?

Ⓐ $x = 3y$ Ⓑ $y = 3x$

Ⓒ $x = 12y$ Ⓓ $y = 12x$

Ⓔ $y = \frac{1}{2}x$

6. *Multiple Choice*   Choose the ratio form of the direct variation model $y = kx$.

Ⓐ $k = \frac{x}{y}$ Ⓑ $k = \frac{-x}{y}$

Ⓒ $k = \frac{y}{x}$ Ⓓ $k = \frac{x}{-y}$

Ⓔ $k = xy$

7. *Multiple Choice*   One year in a dog's life is equivalent to seven years in a human's life. Choose the model that relates a dog's life d to a human's life h.

Ⓐ $d = -7h$ Ⓑ $d = 7h$

Ⓒ $d = \frac{1}{7}h$ Ⓓ $h = \frac{1}{7}d$

Ⓔ $h = -\frac{1}{7}d$

Quantitative Comparison   In Exercises 8–10, choose the statement that is true about the given number.

Ⓐ The number in column A is greater

Ⓑ The number in column B is greater

Ⓒ The two numbers are equal

Ⓓ The relationship cannot be determined from the information given.

	Column A	Column B
8.	The constant of variation in $y = -3x$	The constant of variation in $5x = y$
9.	The value of y when $x = 3$ and $y = -\frac{1}{9}x$	The value of x when $y = -2$ and $y = 2x$
10.	The slope of the line $2y + x = 0$	The constant of variation in $y = -\frac{1}{2}x$

Algebra 1
Standardized Test Practice Workbook

Chapter 4

NAME _____ DATE _____

Standardized Test Practice

For use with pages 241–247

TEST TAKING STRATEGY   Think positively during a test. This will help keep up your confidence and enable you to focus on each question.

1. *Multiple Choice*   What is the slope of a linear equation written in slope-intercept form, $y = mx + b$?
   - (A) y
   - (B) x
   - (C) m
   - (D) b
   - (E) mx

2. *Multiple Choice*   What is the slope of the graph of $y = -3x + 2$?
   - (A) 1
   - (B) -3
   - (C) 2
   - (D) 3
   - (E) -2

3. *Multiple Choice*   What is the slope of the graph of $4x + 3y = 15$?
   - (A) 4
   - (B) -4
   - (C) $-\frac{4}{3}$
   - (D) $\frac{4}{3}$
   - (E) 5

4. *Multiple Choice*   What is the y-intercept of the graph of $y = 5x - 2$?
   - (A) 1
   - (B) 5
   - (C) 2
   - (D) -2
   - (E) -5

5. *Multiple Choice*   What is the y-intercept of the graph of $3y - 2x = 18$?
   - (A) 18
   - (B) $\frac{2}{3}$
   - (C) 6
   - (D) -6
   - (E) $-\frac{2}{3}$

6. *Multiple Choice*   Choose the set of equations which are parallel lines.
   - (A) $x = 2, y = 6$
   - (B) $y = 3x + 6, y = 2x + 6$
   - (C) $y = -5x + 1, y = 5x + 3$
   - (D) $y = -\frac{1}{2}x + 2, y = -\frac{1}{2}x$
   - (E) $y = 6x + 3, y = \frac{1}{6}x + 3$

7. *Multiple Choice*   What is the equation, in slope-intercept form, of the line shown.

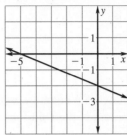

   - (A) $-2x - 5y = 10$
   - (B) $y = -\frac{2}{5}x + 2$
   - (C) $y = -\frac{2}{5}x - 2$
   - (D) $-5x - 2y = 10$
   - (E) $y = \frac{2}{5}x - 2$

8. *Multiple Choice*   You are filling a pot with water. The water level in the pot is rising at a rate of 2 inches per minute. The pot is already 3 inches full. The equation $y = 2x + 3$ models the depth of the water after x minutes. What is the depth of the water after 3 minutes?
   - (A) 3
   - (B) 7
   - (C) 6
   - (D) 8
   - (E) 9

Quantitative Comparison   In Exercises 9–11, choose the statement that is true about the given number.
   - (A) The number in column A is greater
   - (B) The number in column B is greater
   - (C) The two numbers are equal
   - (D) The relationship cannot be determined from the information given.

	Column A	Column B
9.	The slope of $y = -3x + 6$	The slope of $y = -2x + 2$
10.	The slope of $y = -5x$	The slope of $2y = -10x + 15$
11.	The y-intercept of $y = \frac{1}{2}x + 3$	The y-intercept of $y = 13x$

Standardized Test Practice

For use with pages 250–255

TEST TAKING STRATEGY   Before you give up on a question, try to eliminate some of your choices so you can make an educated guess.

1. **Multiple Choice**   Which equation below is written in the form $ax + b = 0$?

   (A) $2x + 3y = 7$ (B) $y = \frac{1}{2}x + 2$

   (C) $3x + 6 = y$ (D) $-2x - 17 = 0$

   (E) $5x - 2y = 0$

2. **Multiple Choice**   Match the equation with the line whose x-intercept is the solution of the equation.

   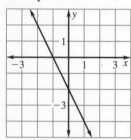

   (A) $3x + 6 = 0$ (B) $x - 1 = 0$

   (C) $1 - x = 0$ (D) $\frac{1}{2}x - \frac{1}{2} = 0$

   (E) $5x + 5 = 0$

3. **Multiple Choice**   Rewrite the equation $7 - 3x = 15$ in $ax + b = 0$ form.

   (A) $-3x = 8$ (B) $-3x + 8 = 0$

   (C) $x = -\frac{8}{3}$ (D) $-3x - 8 = 0$

   (E) $y = -3x - 8$

4. **Multiple Choice**   Rewrite the equation $8 - \frac{2}{3}x = 17 + \frac{1}{3}x$ in $y = ax + b$ form.

   (A) $8 = 17 + x$ (B) $x + 9 = 0$

   (C) $y = x + 9$ (D) $y = \frac{1}{3}x + 9$

   (E) $x = -9$

5. **Multiple Choice**   Solve the equation $-\frac{2}{5}x + 6 = \frac{6}{5}$ algebraically.

   (A) $10\frac{4}{5}$ (B) 12 (C) $-7\frac{1}{5}$

   (D) $-10\frac{4}{5}$ (E) $3\frac{1}{5}$

6. **Multiple Choice**   The graph below shows the solution of which equation?

   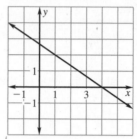

   (A) $3y + 2x = 8$ (B) $4y + 3x = 6$

   (C) $2y + 3x = 8$ (D) $y = \frac{2}{3}x + 4$

   (E) $y = -\frac{1}{2}x + \frac{8}{3}$

7. **Multi-Step Problem**   A store owner is trying to predict next years sales for his store. He developed two models to estimate for an average year and a good year. In the models, x is the number of year since 1994.

   Model 1: $S = 13,850x + 220,000$

   Model 2: $S = 24,750x + 220,000$

   a. Write equations to solve for the year each model predicts sales will be $350,000.

   b. Graph the equation on one coordinate plane. In what year will sales reach $350,000 for each model?

Chapter 4

NAME _____ DATE _____

Standardized Test Practice

For use with pages 256–262

TEST TAKING STRATEGY   **Be aware of how much time you have left, but keep focused on your work.**

1. *Multiple Choice*   Which of the relations below is *not* a function.

Ⓐ
Input	1	2	3	4
Output	2	4	6	8

Ⓑ
Input	0	1	2	3
Output	2	2	2	2

Ⓒ
Input	0	1	1	2
Output	1	2	3	4

Ⓓ
Input	1	2	3	4
Output	2	2	4	4

Ⓔ
Input	1	2	3	4
Output	5	6	7	8

2. *Multiple Choice*   Evaluate the function $f(x) = \frac{3}{2}x + 6$ when $x = 2$.
Ⓐ 9   Ⓑ 6   Ⓒ −9
Ⓓ −6   Ⓔ $-\frac{8}{3}$

3. *Multiple Choice*   Evaluate the function $g(x) = 2x^2 - 7$ for $x = -3$.
Ⓐ 5   Ⓑ −5   Ⓒ −25
Ⓓ 11   Ⓔ −11

4. *Multiple Choice*   Find the slope of the graph of the linear function f if $f(3) = 6$ and $f(-2) = -4$.
Ⓐ $-\frac{2}{5}$ Ⓑ $\frac{1}{2}$ Ⓒ −2
Ⓓ 2   Ⓔ $-\frac{5}{2}$

5. *Multiple Choice*   Find the slope of the graph of the linear function g if $g(-2) = 4$ and $g(2) = 5$.
Ⓐ 0   Ⓑ 9   Ⓒ 4
Ⓓ $-\frac{1}{4}$ Ⓔ $\frac{1}{4}$

6. *Multiple Choice*   The line below matches which function?

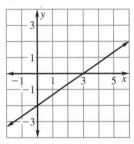

Ⓐ $f(x) = -\frac{2}{3}x - 2$ Ⓑ $f(x) = \frac{2}{3}x - 2$
Ⓒ $f(x) = \frac{2}{3}x + 2$ Ⓓ $f(x) = \frac{2}{3}x + 3$
Ⓔ $f(x) = \frac{2}{3}x - 3$

7. *Multiple Choice*   If it takes 1.3 hours for a stick to float $\frac{3}{4}$ of a mile downstream, choose the correct linear function expressing the distance traveled as a function of time.
Ⓐ $f(t) = \frac{1.3}{0.75}t$ Ⓑ $f(t) = 1.3t$
Ⓒ $f(t) = 0.58t$ Ⓓ $f(t) = 0.75t$
Ⓔ $f(t) = 1.73t$

Quantitative Comparison   In Exercises 8–10, choose the statement that is true about the given number.

Ⓐ   The number in column A is greater
Ⓑ   The number in column B is greater
Ⓒ   The two numbers are equal
Ⓓ   The relationship cannot be determined from the information given.

	Column A	Column B
8.	$f(x) = -5x + 2$ when $x = 3$	$f(x) = -5x + 2$ when $x = -3$
9.	$g(x) = \frac{1}{2}x - 3$ when $x = 4$	$g(x) = -\frac{2}{3}x + 3$ when $x = 9$
10.	$f(x) = x^2 + 3$ when $x = -2$	$f(x) = -x^2 + 3$ when $x = -2$

Chapter 4

NAME _____ DATE _____

Standardized Test Practice

For use with pages 273–278

TEST TAKING STRATEGY   Be aware of how much time you have left, but keep focused on your work.

1. *Multiple Choice*   An equation of the line whose slope is 3 and whose *y*-intercept is 8 is ___?___?

   Ⓐ $y = 8x + 3$ Ⓑ $y = 3x + 8$

   Ⓒ $y = -3x + 8$ Ⓓ $y = -3x - 8$

   Ⓔ $y = 3x - 8$

2. *Multiple Choice*   An equation of the line whose slope is -6 and whose *y*-intercept is 9 is ___?___?

   Ⓐ $y = -9x - 6$ Ⓑ $y = 9x - 6$

   Ⓒ $y = -6x - 9$ Ⓓ $y = 6x - 9$

   Ⓔ $y = -6x + 9$

3. *Multiple Choice*   An equation of the line whose slope is $\frac{1}{2}$ and whose *y*-intercept is -2 is ___?___?

   Ⓐ $y = \frac{1}{2}x - 2$ Ⓑ $y = -\frac{1}{2}x - 2$

   Ⓒ $y = -2x - \frac{1}{2}$ Ⓓ $y = -2x + \frac{1}{2}$

   Ⓔ $y = \frac{1}{2}x + 2$

4. *Multiple Choice*   What is an equation of the line shown in the graph?

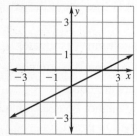

   Ⓐ $y = 2x - 1$ Ⓑ $y = \frac{1}{2}x + 1$

   Ⓒ $y = \frac{1}{2}x - 1$ Ⓓ $y = -\frac{1}{2}x + 1$

   Ⓔ $y = -x + \frac{1}{2}$

5. *Multiple Choice*   Write a set of equations for parallel lines shown in the graph.

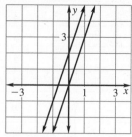

   Ⓐ $y = 3x; y = 3x - 2$

   Ⓑ $y = 3x; y = 3x + 2$

   Ⓒ $y = \frac{1}{3}x; y = \frac{1}{3}x + 2$

   Ⓓ $y = \frac{1}{3}x; y = \frac{1}{3}x - 1$

   Ⓔ $y = 3x; y = 3x - 1$

Quantitative Comparison   In Exercises 6–9, choose the statement below that is true about the given number.

   Ⓐ   The number is column A is greater.

   Ⓑ   The number in column B is greater.

   Ⓒ   The two numbers are equal.

   Ⓓ   The relationship cannot be determined from the given information.

	Column A	Column B
6.	The *y*-intercept of $y = 3x - 5$	The slope of $y = 3x - 5$
7.	The slope of $y = -\frac{1}{2}x - \frac{1}{5}$	The *y*-intercept of $y = -\frac{1}{2}x - \frac{1}{5}$
8.	The *x*-intercept of $y = 5x + 6$	The *y*-intercept of $y = \frac{2}{5}x - \frac{6}{5}$
9.	The *y*-intercept of $y = 3x$	The *x*-intercept of $y = 2x + 1$

LESSON 5.2

Standardized Test Practice

For use with pages 279–284

TEST TAKING STRATEGY   Some questions involve more than one step. Reading too quickly might lead to mistaking the answer to a preliminary step for your final answer.

1. *Multiple Choice*   What is an equation of the line that passes through the point $(3, 2)$ and has a slope of 3?

- (A) $y = 3x - 3$
- (B) $y = 3x + 3$
- (C) $y = 3x - 7$
- (D) $y = 3x + 7$
- (E) $y = 3x + 2$

2. *Multiple Choice*   What is an equation of the line that passes through the point $(-2, 4)$ and has a slope of -5?

- (A) $y = -5x - 14$
- (B) $y = -5x - 6$
- (C) $y = -5x + 18$
- (D) $y = -5x - 18$
- (E) $y = 5x + 18$

3. *Multiple Choice*   What is an equation of the line that passes through the point $(1, -3)$ and has slope of $-\frac{1}{2}$?

- (A) $y = \frac{1}{2}x - \frac{2}{5}$
- (B) $y = \frac{1}{2}x + \frac{2}{5}$
- (C) $y = -\frac{1}{2}x + \frac{5}{2}$
- (D) $y = -\frac{1}{2}x - \frac{5}{2}$
- (E) $y = -\frac{1}{2}x - 3$

4. *Multiple Choice*   Write an equation of the line shown in the graph.

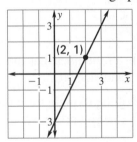

- (A) $y = 2x + 3$
- (B) $y = \frac{1}{2}x + 3$
- (C) $y = 2x - 3$
- (D) $y = \frac{1}{2}x - 3$
- (E) $y = -2x - 3$

5. *Multiple Choice*   What is an equation of the line that is parallel to $y = 3x + 1$ and passes through the point $(2, 6)$?

- (A) $y = 3x + 2$
- (B) $y = 3x$
- (C) $y = 3x + 6$
- (D) $y = 3x + 2$
- (E) $y = -3x$

6. *Multiple Choice*   What is an equation of the line that is parallel to $y = \frac{1}{2}x - 3$ and has an x-intercept of -4?

- (A) $y = \frac{1}{2}x - 4$
- (B) $y = \frac{1}{2}x - 2$
- (C) $y = -\frac{1}{2}x - 2$
- (D) $y = -\frac{1}{2}x + 2$
- (E) $y = \frac{1}{2}x + 2$

Quantitative Comparison   In Exercises 7–9, choose the statement below that is true about the given number.

- (A) The number is column A is greater.
- (B) The number in column B is greater.
- (C) The two numbers are equal.
- (D) The relationship cannot be determined from the given information.

	Column A	Column B
7.	The y-intercept of the line with $m = 2$ and passing through $(1, 3)$	The y-intercept of the line with $m = -1$ and passing through $(3, -5)$
8.	The y-intercept of the line with $m = \frac{1}{2}$ and passing through $(6, -3)$	The y-intercept of the line with $y = 4x + 2$ and passing through $(1, 2)$.
9.	The slope of the line parallel to $y = \frac{1}{5}x$	The y-intercept of the line with $m = \frac{3}{2}$ and passing through $(4, -3)$

LESSON

5.3

Chapter 5

NAME _____ DATE _____

Standardized Test Practice

For use with pages 285–291

TEST TAKING STRATEGY   Go back and check as much of your work as you can.

1. *Multiple Choice*   What is an equation of the line that passes through points $(2, 3)$ and $(-1, 4)$?

Ⓐ $y = 3x - \frac{11}{3}$ Ⓑ $y = -\frac{1}{3}x - \frac{11}{3}$

Ⓒ $y = -3x + \frac{11}{3}$ Ⓓ $y = -\frac{1}{3}x + \frac{11}{3}$

Ⓔ $y = -\frac{1}{3}x + 3$

2. *Multiple Choice*   What is an equation of the line that passes through the points $(-3, 5)$ and $(2, 15)$?

Ⓐ $y = 2x + 11$ Ⓑ $y = 2x - 11$

Ⓒ $y = -2x + 11$ Ⓓ $y = \frac{1}{2}x + 11$

Ⓔ $y = -\frac{1}{2}x - 11$

3. *Multiple Choice*   What is the slope of the line perpendicular to the line $y = \frac{1}{2}x - 6$?

Ⓐ 2   Ⓑ $-\frac{1}{2}$ Ⓒ -2

Ⓓ $\frac{1}{2}$ Ⓔ $\frac{1}{6}$

4. *Multiple Choice*   Write an equation of the line shown in the graph.

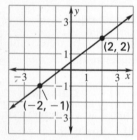

Ⓐ $y = -\frac{3}{4}x + \frac{1}{2}$ Ⓑ $y = \frac{4}{3}x + \frac{1}{2}$

Ⓒ $y = \frac{3}{4}x - \frac{1}{2}$ Ⓓ $y = \frac{4}{3}x - \frac{1}{2}$

Ⓔ $y = \frac{3}{4}x + \frac{1}{2}$

5. *Multiple Choice*   Which set of lines are perpendicular?

Ⓐ $y = 2x + 3; y = -2x + 6$

Ⓑ $y = \frac{1}{3}x; y = -3x + 2$

Ⓒ $y = \frac{2}{3}x + \frac{1}{2}; y = \frac{2}{3}x - 2$

Ⓓ $y = -6x + 1; y = -\frac{1}{6}x - 1$

Ⓔ $y = 3; y = x + 1$

6. *Multiple Choice*   Which lines are perpendicular?

Line d passes through $(2, 4)$ and $(-1, 6)$.

Line e passes through $(-3, -2)$ and $(5, 8)$.

Line f passes through $(2, 10)$ and $(7, 6)$.

Ⓐ Lines d and e

Ⓑ Lines d and f

Ⓒ Lines e and f

Ⓓ All three are perpendicular

Ⓔ None of these

7. *Multiple Choice*   Which is an equation of a line perpendicular to the line $y = -\frac{3}{2}x + \frac{1}{2}$?

Ⓐ $y = -\frac{2}{3}x + 6$ Ⓑ $y = \frac{2}{3}x + 3$

Ⓒ $y = \frac{3}{2}x - \frac{1}{2}$ Ⓓ $y = -\frac{2}{3}x - \frac{1}{2}$

Ⓔ $y = -\frac{3}{2}x - 2$

8. *Multi-Step Problem*   The triangle shown is a right triangle.

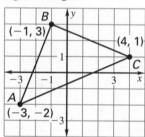

a. Find the equations of the lines that form side $\overline{AB}$, side $\overline{BC}$, and side $\overline{AC}$.

b. Which two sides of the triangle form the right angle? How can you prove it?

LESSON 5.4

Standardized Test Practice

For use with pages 292–298

TEST TAKING STRATEGY   Avoid spending too much time on one question. Skip questions that are too difficult for you and spend no more than a few minutes on each question.

Multiple Choice   In Exercises 1–6 refer to the scatter plots below.

I.
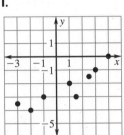

II.
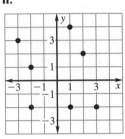

III.
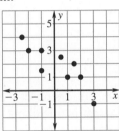

IV.
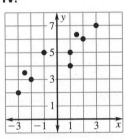

1. Which plot has a positive correlation?

  A I   **B** II   **C** IV

  D I and III   **E** I and IV

2. Which plot has a negative correlation?

  A I   **B** II   **C** III

  D IV   **E** I and IV

3. Which plot has no correlation?

  A I   **B** II   **C** III

  D IV   **E** I and IV

4. Which equation would best fit Graph IV?

  A $y = \frac{3}{4}x + \frac{9}{2}$ **B** $y = -\frac{3}{4}x + \frac{9}{2}$

  C $y = \frac{3}{4}x - \frac{6}{2}$ **D** $y = \frac{3}{4}x + 3$

  E $y = \frac{1}{2}x + 2$

5. Which equation would best fit Graph III

  A $y = x + 2$ **B** $y = x + 4$

  C $y = \frac{1}{2}x + 2$ **D** $y = -x + 2$

  E $y = -\frac{1}{2}x + 3$

6. Which equation would best fit Graph I?

  A $y = -\frac{2}{3}x - \frac{5}{2}$ **B** $y = -2x - 1$

  C $y = \frac{2}{3}x - \frac{5}{2}$ **D** $y = \frac{2}{3}x + \frac{5}{2}$

  E $y = 2x - 1$

7. *Multi-Step Problem*   The average price of a pound of large gulf shrimp is shown in the table.

Year	1992	1993	1994	1995
Dollars / lb	8.40	9.10	9.80	10.50

Year	1996	1997	1998
Dollars / lb	11.20	13.50	15.90

  a. Draw a scatter plot of the price of shrimp letting x represent the number of years since 1992.

  b. Find the equation of the line that best fits the data.

  c. Use the equation to approximate the price per pound in 2001.

Standardized Test Practice

For use with pages 300–306

TEST TAKING STRATEGY   **As soon as the testing begins, start working. Keep moving and stay focused on the test.**

1. *Multiple Choice*   What is the value of x_1 in the equation $y - 6 = \frac{1}{2}(x + 2)$?

   Ⓐ 6   Ⓑ -6 Ⓒ 2

   Ⓓ -2 Ⓔ $\frac{1}{2}$

2. *Multiple Choice*   Which equation in point-slope form passes through the point (2, 3) and has a slope of $-\frac{1}{3}$?

   Ⓐ $y + 3 = -\frac{1}{3}(x - 2)$

   Ⓑ $y - 3 = -\frac{1}{3}(x - 2)$

   Ⓒ $y - 2 = -\frac{1}{3}(x - 3)$

   Ⓓ $y + 2 = -\frac{1}{3}(x + 3)$

   Ⓔ $y = -\frac{1}{3}x + \frac{7}{3}$

3. *Multiple Choice*   Which equation in point-slope form passes through the point $(-1, 6)$ and has slope of $\frac{2}{3}$?

   Ⓐ $y - 6 = \frac{2}{3}(x + 1)$ Ⓑ $y - 1 = \frac{2}{3}(x + 6)$

   Ⓒ $y + 6 = \frac{2}{3}(x - 1)$ Ⓓ $y - 6 = \frac{2}{3}(x - 1)$

   Ⓔ $y + 1 = \frac{2}{3}(x - 6)$

4. *Multiple Choice*   Which equation in point-slope form passes through the points (3, 5) and (2, 8)?

   Ⓐ $y - 8 = -3(x - 3)$

   Ⓑ $y - 5 = -3(x - 3)$

   Ⓒ $y - 5 = 3(x - 3)$

   Ⓓ $y - 5 = -\frac{1}{3}(x - 3)$

   Ⓔ $y - 8 = -\frac{1}{3}(x + 2)$

5. *Multiple Choice*   Which equation in point-slope form passes through the points $(-2, 6)$ and $(3, -1)$?

   Ⓐ $y + 1 = \frac{7}{5}(x - 3)$

   Ⓑ $y + 1 = -\frac{5}{7}(x - 3)$

   Ⓒ $y + 2 = -\frac{7}{5}(x - 6)$

   Ⓓ $y - 1 = -\frac{7}{5}(x + 3)$

   Ⓔ $y + 1 = -\frac{7}{5}(x - 3)$

6. *Multiple Choice*   Which is the equation for the line shown in the graph?

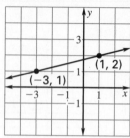

   Ⓐ $y + 3 = -\frac{1}{4}(x - 1)$

   Ⓑ $y - 2 = -4(x - 1)$

   Ⓒ $y - 1 = \frac{1}{4}(x + 3)$

   Ⓓ $-y - 2 = 4(x - 1)$

   Ⓔ $-y - 1 = -\frac{1}{4}(x + 3)$

Quantitative Comparison   In Exercises 7–9, choose the statement below that is true about the given number.

   Ⓐ   The number is column A is greater.

   Ⓑ   The number in column B is greater.

   Ⓒ   The two numbers are equal.

   Ⓓ   The relationship cannot be determined from the given information.

	Column A	Column B
7.	The value of m in $y + 3 = \frac{1}{2}(x - 2)$	The value of x_1 in $y - 6 = \frac{1}{2}(x - 2)$
8.	The value of y_1 in $y + 3 = -2(x - 3)$	The value of m in the line passing through (1, 8) and (2, 3)
9.	The value of m in the line passing through $(-1, -5)$ and $(2, -3)$	The value of y_1 in $y - 6 = 7(x - 8)$

LESSON 5.6

Standardized Test Practice

For use with pages 308–314

TEST TAKING STRATEGY   Think positively during a test. This will help keep up your confidence and enable you to focus on each question.

1. *Multiple Choice*   Write $y = \frac{3}{7}x - 2$ in standard form with integer coefficients.

   Ⓐ $7y = 3x - 14$ Ⓑ $-3x + 7y = -2$
   Ⓒ $-3x + 7y = 14$ Ⓓ $3x - 7y = 14$
   Ⓔ $3x - 7y = -14$

2. *Multiple Choice*   What is the standard form of an equation of the line that passes through the point $(-1, 5)$ and has a slope of $\frac{1}{2}$?

   Ⓐ $x - 2y = -9$ Ⓑ $x + 2y = 9$
   Ⓒ $x + 2y = -11$ Ⓓ $x + 2y = 11$
   Ⓔ $x - 2y = -11$

3. *Multiple Choice*   What is the standard form of an equation of the line that passes through the point $(-8, -2)$ and has a slope of -5?

   Ⓐ $5x + y = 42$ Ⓑ $5x + y = -42$
   Ⓒ $5x + y = -38$ Ⓓ $5x - y = 38$
   Ⓔ $5x + y = -18$

4. *Multiple Choice*   What is the standard form of an equation of the line that passes through the points $(-1, 4)$ and $(-7, -5)$?

   Ⓐ $y - 4 = \frac{3}{2}(x + 1)$
   Ⓑ $3x + 2y = 11$
   Ⓒ $-3x + 2y = 11$
   Ⓓ $-3x + 2y = -5$
   Ⓔ $3x + 2y = -5$

5. *Multiple Choice*   What is the standard form of an equation of the line that passes through the points $(3, 0)$ and $(6, -8)$?

   Ⓐ $8x + 3y = 24$ Ⓑ $y = -\frac{8}{3}x + 8$
   Ⓒ $8x + 3y = 8$ Ⓓ $3x + 8y = 64$
   Ⓔ $8x + 3y = -24$

6. *Multiple Choice*   What is the standard form of an equation of the horizontal line that passes through the point $(2, -8)$?

   Ⓐ $y = 2$ Ⓑ $x = 2$
   Ⓒ $y = -8$ Ⓓ $x = -8$
   Ⓔ $y = 0$

7. *Multiple Choice*   What is the standard form of an equation of the vertical line in the graph?

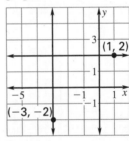

   Ⓐ $y = -2$ Ⓑ $x = -3$
   Ⓒ $y = 2$ Ⓓ $x = 1$
   Ⓔ $y = -3$

Quantitative Comparison   In Exercises 8–10, choose the statement below that is true about the given number.

   Ⓐ The number is column A is greater.

   Ⓑ The number in column B is greater.

   Ⓒ The two numbers are equal.

   Ⓓ The relationship cannot be determined from the given information.

	Column A	Column B
8.	The constant term in $3x + 2y = 7$	A in the standard form of $5x + 8y = 7$
9.	The x-intercept of $3x - 2y = -10$	The y-intercept of $2x + 5y = 6$
10.	The slope of $x = -6$	The slope of $12x + 3y = 16$

NAME _____ DATE _____

Standardized Test Practice

For use with pages 316–322

TEST TAKING STRATEGY   Read all of the answer choices before deciding which is the correct one.

Multiple Choice   In Exercises 1 and 2 refer to the scatter plots below.

I.

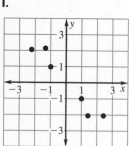

II.

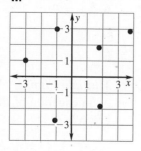

III.

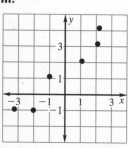

IV.

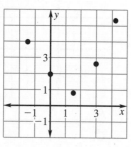

1. Which graph could be modeled with a linear model?

  Ⓐ I   Ⓑ II   Ⓒ III

  Ⓓ I and III   Ⓔ I, III and IV

2. Which linear model could be used to model graph III?

  Ⓐ $y = -x + 1$ Ⓑ $y = x + 1$

  Ⓒ $y = x - 1$ Ⓓ $y = x + 3$

  Ⓔ $y = 3x + 1$

3. *Multiple Choice*   If you are using linear extrapolation, then you are estimating a data point __?__.

  Ⓐ above the data points given

  Ⓑ below the data points given

  Ⓒ to the right of the data points given

  Ⓓ between the given data points

  Ⓔ none of these

4. *Multiple Choice*   If you are using linear interpolation, then you are estimating a data point __?__.

  Ⓐ above the data points given

  Ⓑ below the data points given

  Ⓒ to the right of the data points given

  Ⓓ between the given data points

  Ⓔ none of these

5. *Multi-Step Problem*   The table below shows the average weekly hours spent watching TV and using the Internet by children.

Year	1990	1992	1994	1996	1998
TV	20	19	18	15	14
Internet	2	5	7	10	13

  a. Make a scatter plot of the number of hours spent watching TV in terms of the year t. Let t be the number of years since 1990.

  b. Make a scatter plot of the number of hours spent using the Internet in terms of the year t. Let t be the number of years since 1990.

  c. Write a linear model for hours spent watching TV and hours using the Internet.

  d. Use the linear model for the hours spent on the Internet to estimate the number of hours in 2001. Is this linear interpolation or extrapolation?

  e. Use the model for hours spent watching TV to estimate the hours spent watching TV in 1995. Is this linear interpolation or extrapolation?

Algebra 1
Standardized Test Practice Workbook

Standardized Test Practice

For use with pages 334–339

TEST TAKING STRATEGY   **Learn as much as you can about a test ahead of time, such as the types of questions and the topics that the test will cover.**

1. *Multiple Choice* Describe the solution of $2x \leq 14$.

  Ⓐ All real numbers less than 7

  Ⓑ All real numbers less than or equal to 7

  Ⓒ All real numbers greater than or equal to 7

  Ⓓ All real numbers less than 12

  Ⓔ None of these

2. *Multiple Choice* Describe the solution of $x - 8 > 3$.

  Ⓐ All real numbers greater than -5

  Ⓑ All real numbers less than -5

  Ⓒ All real numbers greater than 11

  Ⓓ All real numbers less than 11

  Ⓔ None of these

3. *Multiple Choice* Choose the solution of $-5 + x > 10$.

  Ⓐ $x > 5$ Ⓑ $x > -5$

  Ⓒ $x > 15$ Ⓓ $x < 15$

  Ⓔ $x < 5$

4. *Multiple Choice* Choose the solution of $-\frac{x}{7} \geq 3$.

  Ⓐ $x \geq -4$ Ⓑ $x \geq -21$

  Ⓒ $x \leq -21$ Ⓓ $x \leq -\frac{3}{7}$

  Ⓔ $x \leq -4$

5. *Multiple Choice* The coldest temperature recorded in your city was $-6°C$. Choose the inequality describing all of the recorded temperatures in your city.

  Ⓐ $t < -6$ Ⓑ $t > -6$

  Ⓒ $t \leq -6$ Ⓓ $t \geq -6$

  Ⓔ None of these

6. *Multiple Choice* Which graph represents the solution of $x + 7 > 3$?

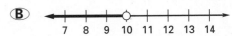

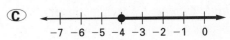

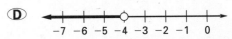

7. *Multiple Choice* Which graph represents the solution of $-3x \leq 6$?

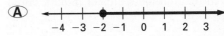

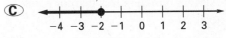

Quantitative Comparison   In Exercises 8–10, choose the statement that is true about the given number.

  Ⓐ The number in column A is greater.

  Ⓑ The number in column B is greater.

  Ⓒ The two numbers are equal.

  Ⓓ The relationship cannot be determined from the given information.

	Column A	Column B
8.	$x < 6$	$x \geq 6$
9.	$x - 10 \geq -3$	$2x < 10$
10.	$-\frac{x}{3} \leq 5$	$x + 7 \geq -8$

NAME _____ DATE _____

Standardized Test Practice

For use with pages 340–345

TEST TAKING STRATEGY Be aware of how much time you have left, but keep focused on your work.

1. *Multiple Choice* Which inequality is equivalent to $3x - 2 \leq 7$?

   (A) $x \geq 3$ (B) $x \leq 3$

   (C) $x \leq \frac{5}{3}$ (D) $x \geq \frac{5}{3}$

   (E) $x \leq 6$

2. *Multiple Choice* Which inequality is equivalent to $5 - 8x \geq -11$?

   (A) $x \leq 2$ (B) $x \geq 2$

   (C) $x \leq -2$ (D) $x \leq \frac{3}{4}$

   (E) $x \geq \frac{3}{4}$

3. *Multiple Choice* Describe the solution of the inequality $-\frac{1}{2}x - 4 \leq 6$.

   (A) All real numbers greater than or equal to -20

   (B) All real numbers greater than or equal to 20

   (C) All real numbers less than or equal to -20

   (D) All real numbers less than or equal to -5

   (E) All real numbers greater than or equal to -5

4. *Multiple Choice* Which inequality is equivalent to $x - 5 \leq 3x + 7$?

   (A) $x \geq 6$ (B) $x \geq -1$

   (C) $x \leq -6$ (D) $x \geq -6$

   (E) $x \leq -1$

5. *Multiple Choice* Which inequality is equivalent to $6 - x > 4x + 9$?

   (A) $x < 3$ (B) $x > 3$

   (C) $x < -\frac{3}{5}$ (D) $x > -\frac{3}{5}$

   (E) $x < -3$

6. *Multiple Choice* It costs $20 to golf at a local course. A season pass costs $250. Which inequality represents the number of games you need to play to justify buying a season pass?

   (A) $x \geq 12$ (B) $x > 12$

   (C) $x \geq 13$ (D) $x \leq 12$

   (E) None of these

7. *Multi-Step Problem* You are working on your school's yearbook. It takes at least 3 hours to design and layout a page.

   a. Write an inequality that describes the number of hours it would take to layout a 220 page yearbook.

   b. You work 4 hours per week. Write an inequality describing the number of weeks it would take you to complete the yearbook if you were working alone.

   c. Write an inequality to describe the number of people that would be necessary to complete the yearbook in 20 weeks or less if each person averages 5 hours of work per week.

Chapter 6

NAME _____ DATE _____

Standardized Test Practice

For use with pages 346–352

TEST TAKING STRATEGY   Some questions involve more than one step. Reading too quickly might lead to mistaking the answer to a preliminary step for your final answer.

1. *Multiple Choice*   Which inequality represents all real numbers greater than 4 and less than or equal to 7?

   Ⓐ $4 < x \le 7$ Ⓑ $4 \le x \le 7$

   Ⓒ $4 < x < 7$ Ⓓ $4 > x$ or $x \le 7$

   Ⓔ $4 < x$ or $x \ge 7$

2. *Multiple Choice*   Describe the solution of the compound inequality $-2x + 5 < 9$ or $-5x + 2 > 17$.

   Ⓐ All real numbers less than -3 and greater than 2

   Ⓑ All real numbers less than 3 and greater than -2

   Ⓒ All real numbers less than 3 and greater than 2

   Ⓓ All real numbers less than -3 or greater than -2

   Ⓔ All real numbers less than $-\frac{19}{5}$ and greater than -7

3. *Multiple Choice*   Describe the solution of the inequality $8 \le x - 3 < 12$.

   Ⓐ All real numbers less than 9 and greater than or equal to 5

   Ⓑ All real numbers less than 15 or greater than or equal to 11

   Ⓒ All real numbers less than 15 and greater than or equal to 11

   Ⓓ All real numbers less than 9 or greater than or equal to 5

   Ⓔ None of these

4. *Multiple Choice*   Choose the solution of the inequality $6 \le -2x < 14$.

   Ⓐ $-7 < x \le -3$ Ⓑ $-3 \le x < -7$

   Ⓒ $-3 \ge x > -7$ Ⓓ A and C

   Ⓔ A and B

5. *Multiple Choice*   Which graph represents the solution of $-3 \le 4x + 1 < 9$?

   Ⓐ

   Ⓑ

   Ⓒ

   Ⓓ

   Ⓔ

6. *Multiple Choice*   Which graph represents the solution of the inequality $4x + 1 < 13$ or $3x - 8 \ge 10$.

   Ⓐ

   Ⓑ

   Ⓒ

   Ⓓ

   Ⓔ None of these

Quantitative Comparison   In Exercises 7–9, choose the statement below that is true about the given number.

   Ⓐ The number in column A is greater.

   Ⓑ The number in column B is greater.

   Ⓒ The two numbers are equal.

   Ⓓ The relationship cannot be determined from the information given.

	Column A	*Column B*
7.	$3 \le x < 8$	$x \ge -2$ and $x < 3$
8.	$2 \le x + 3 \le 18$	$x \le 14$ or $x > 3$
9.	$8 < -2x + 4 \le 12$	$3x < -6$ and $-\frac{x}{2} \le 2$

NAME _____ DATE _____

Standardized Test Practice

For use with pages 353–358

TEST TAKING STRATEGY   **Work as fast as you can through the easier problems, but not so fast that you are careless.**

1. *Multiple Choice*   Which numbers are solutions to the absolute-value equation $|x + 2| - 3 = 8$?

  (A) 6 and -10 (B) 9 and -13

  (C) 9 and -9 (D) -9 and -13

  (E) 3 and -7

2. *Multiple Choice*   Which numbers are solutions to the absolute-value equation $12 + |3x - 1| = 19$?

  (A) 2 and -2 (B) $\frac{8}{3}$ and $-\frac{8}{3}$

  (C) $\frac{32}{3}$ and -10 (D) $\frac{8}{3}$ and -2

  (E) 10 and -10

3. *Multiple Choice*   Which graph represents the solution of $|3x - 6| \geq 9$?

  (A)

  (B)

  (C)

  (D)

  (E)

4. *Multiple Choice*   Which graph represents the solution of $|2x - 7| + 3 < 12$?

  (A)

  (B)

  (C)

  (D)

  (E)

5. *Multiple Choice*   Which graph represents the solution of $|5x - 2| - 8 \geq 9$?

  (A)

  (B)

  (C)

  (D)

  (E)

6. *Multiple Choice*   Your soccer team averages between 3 and 9 goals per game. Choose the absolute-value inequality describing the average number of goals per game.

  (A) $|x - 3| \leq 9$ (B) $|x - 6| \leq 9$

  (C) $|x - 6| \leq 3$ (D) $|x - 3| \leq 3$

  (E) $|x + 3| \leq 6$

Quantitative Comparison   In Exercises 7–10, choose the statement below that is true about the given number.

  (A) The number in column A is greater.

  (B) The number in column B is greater.

  (C) The two numbers are equal.

  (D) The relationship cannot be determined from the given information.

	Column A	Column B
7.	$|x| = 6$	$|x| = 0$
8.	$|x + 2| < 1$	$|x| \leq 1$
9.	$|x - 7| < 2$	$|2x + 3| < 7$
10.	$|4x - 2| < 6$	$|2x - 1| - 1 < 2$

NAME _____ DATE _____

Standardized Test Practice

For use with pages 360–366

TEST TAKING STRATEGY   **If you can, check your answer using a different method than you used originally to avoid making the same mistake twice**

1. *Multiple Choice*   Which point is a solution of $3x - 5y \geq 8$?

   A $(1, 1)$ **B** $(4, 1)$

   C $(-2, -2)$ **D** $(2, -1)$

   E $(-1, 1)$

2. *Multiple Choice*   Which point is a solution of $4x - 8y < 14$?

   A $(0, 2)$ **B** $(3, -1)$

   C $(0, -2)$ **D** $(5, -2)$

   E $(-3, -4)$

3. *Multiple Choice*   Which point is not a solution of $11x - 6 < y$?

   A $(-1, 3)$ **B** $(2, 10)$

   C $(0, 3)$ **D** $(-4, -1)$

   E $(1, 7)$

4. *Multiple Choice*   Choose the inequality whose solution is shown in the graph.

   A $x > -1$

   B $y > -1$

   C $x \geq -1$

   D $y \geq -1$

   E $x - 1 \geq -1$

   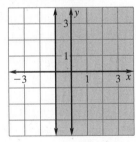

5. *Multiple Choice*   Choose the inequality whose solution is shown in the graph.

   A $y < -3x + 2$

   B $y > -3x + 2$

   C $y > 3x + 2$

   D $y \geq -3x + 2$

   E $y \leq -3x + 2$

   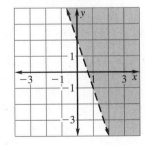

6. *Multiple Choice*   Choose the ratio form of the direct variation model $y = kx$.

   A $y > \frac{2}{3}x - 1$

   B $y < \frac{2}{3}x - 1$

   C $y \leq -\frac{2}{3}x - 1$

   D $y \leq -\frac{2}{3}x + 1$

   E $y \geq \frac{2}{3}x - 1$

   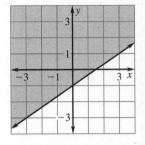

7. *Multi-Step Problem*   You are in charge of a dinner-dance for school. You have a budget of $500 for dinners. There are 3 choices of meals: a fish dinner for $8.50, a steak dinner for $10.50, and a vegetarian dinner for $7.50. Two people order the vegetarian meal.

   a. Write an inequality to model the different combinations of fish and steak dinners that might be purchased.

   b. Graph the inequality.

   c. If only 20 fish dinners are available, how many steak dinners can be purchased?

   d. Does every point on the graph represent a reasonable real-life solution?

NAME _____ DATE _____

Standardized Test Practice

For use with pages 368–374

TEST TAKING STRATEGY   Avoid spending too much time on one question. Skip questions that are too difficult for you, and spend no more than a few minutes on each question.

Multiple Choice   In Exercises 1–5, refer to the stem-and-leaf plots below.

1	0 3 1 6 0
2	2 4 3
3	5 0 1 3 2 2 3
4	5 6 1 1 3 5
5	2 3 1 0

Male ages

1	1 3 0
2	6 3 2 1 3
3	7 0 1 3 0
4	1 2 2
5	5 3 2 2

Female ages

1. How many males were in their thirties?
   - Ⓐ 5
   - Ⓑ 6
   - Ⓒ 7
   - Ⓓ 2
   - Ⓔ 1

2. What percent of females were older than 20?
   - Ⓐ 15%
   - Ⓑ 20%
   - Ⓒ 60%
   - Ⓓ 80%
   - Ⓔ 85%

3. How many people were exactly forty-one years old?
   - Ⓐ 1
   - Ⓑ 2
   - Ⓒ 3
   - Ⓓ 6
   - Ⓔ 9

4. How many males were counted for the plot?
   - Ⓐ 20
   - Ⓑ 45
   - Ⓒ 25
   - Ⓓ 30
   - Ⓔ 55

5. What percent of people counted were under 30 years old?
   - Ⓐ 40%
   - Ⓑ 32%
   - Ⓒ 18%
   - Ⓓ 36%
   - Ⓔ 64%

6. *Multiple Choice*   Your class wants to know the most typical age of a student in your class. It collects the ages of all students and the teacher. Which measures of central tendency would give the best typical student age?
   - Ⓐ mean
   - Ⓑ median
   - Ⓒ mode
   - Ⓓ mean or median
   - Ⓔ median or mode

7. *Multiple Choice*   Your test scores in algebra so far are 78, 85, 72, and 81. If you want to average at least an 80, what score do you need on the next test?
   - Ⓐ 80
   - Ⓑ 84
   - Ⓒ 88
   - Ⓓ 82
   - Ⓔ 90

Quantitative Comparison   In Exercises 8–11, choose the statement below that is true about the given number.
- Ⓐ The number in column A is greater.
- Ⓑ The number in column B is greater.
- Ⓒ The two numbers are equal.
- Ⓓ The relationship cannot be determined from the given information.

	Column A	Column B
8.	The mean of 4, 6, 3, 7, 8	The mean of 9, 6, 2, 6, 4
9.	The median of 3, 5, 2, 5, 8	The mode of 3, 5, 2, 5, 8
10.	The mean of 12, 5, 5, 7, 8, 9	The median of 3, 4, 7, 10, 12, 14
11.	The median of 1, 3, 1, 5, 7, 4	The mode of 3, 5, 7, 6, 2

NAME _____ DATE _____

Standardized Test Practice

For use with pages 375–381

TEST TAKING STRATEGY   Go back and check as much of your work as you like.

1. *Multiple Choice*   Find the first quartile of the collection of numbers: 8, 3, 6, 1, 4, 5, 6, 8, 2, 9, 1, 3.

Ⓐ 1   Ⓑ 2.5   Ⓒ 4.5

Ⓓ 7   Ⓔ 3.5

2. *Multiple Choice*   Find the first quartile of the collection of numbers: 1, 3, 4, 9, 11, 6, 5, 6.

Ⓐ 3.5   Ⓑ 7.5   Ⓒ 5.5

Ⓓ 6.5   Ⓔ 10

3. *Multiple Choice*   Find the first quartile of the collection of numbers: 14, 15, 12, 6, 2, 10, 11, 8, 3, 7, 14, 8.

Ⓐ 6.5   Ⓑ 13   Ⓒ 9.5

Ⓓ 9   Ⓔ 10.5

4. *Multiple Choice*   The third quartile is __?__ .

Ⓐ the median of the data set.

Ⓑ the median of the lower half of data.

Ⓒ the median of the upper half of data.

Ⓓ the mean of the lower half of data.

Ⓔ the mean of the upper half of data

5. *Multiple Choice*   Which set of numbers is represented by the box-and-whisker plot?

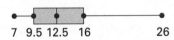

7 9.5 12.5 16   26

Ⓐ 6, 24, 1, 8, 2, 14, 3, 11

Ⓑ 8, 25, 2, 28, 12, 25, 11, 28

Ⓒ 12, 13, 8, 10, 13, 11, 17, 9, 8, 13

Ⓓ 6, 14, 10, 8, 11, 15, 17, 7, 26

Ⓔ 17, 25, 13, 12, 26, 15, 11, 9, 7, 13, 8, 10

6. *Multiple Choice*   In the box-and-whisker plot from Exercise 5, what percent of the data set fell between 9.5 and 16?

Ⓐ 50%   Ⓑ 25%   Ⓒ 75%

Ⓓ 66%   Ⓔ 33%

7. *Multiple Choice*   In the box-and-whisker plot from Exercise 5, what percent of the data fell below 16?

Ⓐ 50%   Ⓑ 25%   Ⓒ 75%

Ⓓ 66%   Ⓔ 33%

8. *Multi-Step Problem*   A school keeps a record of drinks purchased from their cafeteria. The data in the table show the average daily purchases broken down into juice and milk purchases.

Juice	25	17	27	21	23	21	16	26
Milk	10	16	9	15	18	19	22	11

a. Draw a box-and-whisker plot for the juice purchases.

b. Draw a box-and-whisker plot for the milk purchases.

c. What is the median for juice purchases? Milk purchases?

d. *Writing*   How do the data sets differ?

NAME _____ DATE _____

Cumulative Standardized Test Practice

For use after Chapters 1–6

1. *Multiple Choice* Evaluate $\frac{2}{3}x$ when $x = \frac{9}{4}$.

(A) $\frac{11}{7}$ (B) $\frac{3}{2}$ (C) $\frac{2}{3}$

(D) $\frac{18}{7}$ (E) $\frac{4}{9}$

2. *Multiple Choice* Evaluate the expression $(x - y)^3$ when $x = 3$ and $y = -2$.

(A) 1   (B) -1 (C) 125

(D) 25   (E) -125

3. *Multiple Choice* Evaluate $(-2x)^3$ when $x = 3$.

(A) -216 (B) -54 (C) 125

(D) -125 (E) 216

4. *Multiple Choice* What is the value of $\dfrac{3 \cdot 4 + 6^2}{16 - 3^2 \cdot 2}$?

(A) 24   (B) -12 (C) $\frac{13}{4}$

(D) $\frac{48}{5}$ (E) -24

Quantitative Comparison In Exercises 5–7, choose the statement below that is true about the given number.

(A) The number in column A is greater.

(B) The number in column B is greater.

(C) The two numbers are equal.

(D) The relationship cannot be determined from the information given.

	Column A	Column B
5.	$8x - 3 = 29$	$18 = \frac{1}{2}x + 12$
6.	Nine times a number decreased by 8 is 19.	Eight times a number is three squared plus seven.
7.	$y = 3x^2 - 8$ when $x = -2$	$y = 118 - 4x^2$ when $x = 9$

8. *Multiple Choice* Evaluate the expression $-|-6| + 3$.

(A) 3   (B) 9   (C) 0

(D) -3 (E) -9

9. *Multiple Choice* Evaluate the expression $7 + x + (-2)$ when $x = -8$.

(A) 13   (B) 17   (C) -3

(D) -13 (E) 3

10. *Multiple Choice* Evaluate the expression $9x - (2x)^3 + x$ when $x = -2$.

(A) -44 (B) -48 (C) -78

(D) 80   (E) 44

11. *Multiple Choice* Find the difference.

$$\begin{bmatrix} 2 & 3 \\ -1 & 6 \\ 5 & -4 \end{bmatrix} - \begin{bmatrix} 7 & -8 \\ 6 & 3 \\ 2 & -1 \end{bmatrix} = \underline{\ ?\ }$$

(A) $\begin{bmatrix} 9 & -5 \\ 5 & 9 \\ 7 & -5 \end{bmatrix}$ (B) $\begin{bmatrix} -5 & 11 \\ -7 & 3 \\ 3 & -3 \end{bmatrix}$

(C) $\begin{bmatrix} 5 & 11 \\ 5 & 3 \\ 3 & -3 \end{bmatrix}$ (D) $\begin{bmatrix} -5 & -5 \\ -7 & 3 \\ 3 & -5 \end{bmatrix}$

(E) None of these

12. *Multiple Choice* Simplify the equation $8x^2 - x(x - 7) - 2x$.

(A) $9x^2 - 9x$ (B) $7x^2 - 2x - 7$

(C) $9x^2 + 5x$ (D) $7x^2 - 5x$

(E) $7x^2 + 5x$

13. *Multiple Choice* Simplify $-75x^5 \div \dfrac{3x^2}{5}$.

(A) $-30x^2$ (B) $-45x^3$

(C) $-45x^7$ (D) $-125x^3$ (E) $30x^3$

14. *Multiple Choice* There are 13 red balls, 20 blue balls, 15 green balls, and 5 yellow balls in a ball pit. What are the odds of randomly choosing a green ball?

Ⓐ $\frac{13}{53}$ Ⓑ $\frac{15}{53}$ Ⓒ $\frac{15}{38}$

Ⓓ $\frac{13}{38}$ Ⓔ $\frac{38}{13}$

15. *Multiple Choice* Find the missing lengths if the perimeter is 21.

Ⓐ 2, 4

Ⓑ 3, 6

Ⓒ 5, 10

Ⓓ 6, 12   Ⓔ 4, 8

Quantitative Comparison In Exercises 16–18, choose the statement below that is true about the given number.

Ⓐ The value of column A is greater.

Ⓑ The value of column B is greater.

Ⓒ The two columns are equal.

Ⓓ The relationship cannot be determined from the information given.

	Column A	Column B
16.	$2x + |-3| = 17$	$-|-7| + 3x = 19$
17.	$-12 = \frac{x}{4}$	$-\frac{2}{3}x = 32$
18.	$4x - 2(x - 3) = 17$	$\frac{1}{2}(x - 4) = 12$

19. *Multiple Choice* Solve $5x - 6 = 4x + 3$.

Ⓐ 9   Ⓑ 8   Ⓒ 7

Ⓓ 6   Ⓔ 5

20. *Multiple Choice* Solve the equation $\frac{1}{3}(9x - 15) - 2 = 2x + 5$.

Ⓐ 12   Ⓑ 8   Ⓒ 22

Ⓓ −22   Ⓔ −8

21. *Multiple Choice* Solve the equation $\frac{1}{4}(4x - 4) = \frac{1}{5}(5x - 10)$.

Ⓐ 0   Ⓑ 2   Ⓒ −3

Ⓓ 1   Ⓔ no solution

22. *Multi-Step Problem* Your cable TV company charges $32 a month for basic cable and one premium channel. Additional premium channels cost $12 per month.

   a. Let x represent the number of premium channels you subscribe to. Write an expression that would model the total cost of your cable bill.

   b. Your bill came to $56 last month. How many premium channels did you order?

   c. *Critical Thinking* If your total bill also includes 6% sales tax, how could you figure out how many premium channels you ordered?

23. *Multiple Choice* Solve the equation $2.3(-1.2x + 8.2) = 2.1x$. Round to the nearest tenth.

Ⓐ −28.6   Ⓑ 28.6   Ⓒ 3.8

Ⓓ 3.9   Ⓔ −3.9

24. *Multiple Choice* How much money would you have to invest to earn $208 dollars in interest in 5 years at a rate of 8%? (*Hint:* $I = Prt$)

Ⓐ $480   Ⓑ $550   Ⓒ $630

Ⓓ $570   Ⓔ $520

25. *Multiple Choice* Rewrite the equation $5x + 3y = 12$ so that y is a function of x.

Ⓐ $y = -\frac{5}{3}x + 4$ Ⓑ $y = \frac{5}{3}x + 4$

Ⓒ $y = \frac{3}{5}x + 4$ Ⓓ $y = 2x + 9$

Ⓔ $y = -2x + 9$

26. *Multiple Choice* Which quadrant of the coordinate plane would point $(-3, 6)$ be in?

Ⓐ I   Ⓑ II   Ⓒ III

Ⓓ IV   Ⓔ II or IV

NAME _____ DATE _____

Cumulative Standardized Test Practice

Chapter 6

27. *Multiple Choice*   Choose the ordered pair that is a solution of $3x - 2y = 8$.

Ⓐ $(-4, -2)$ Ⓑ $(2, 1)$ Ⓒ $(8, 6)$

Ⓓ $(2, -1)$ Ⓔ $(-2, -8)$

28. *Multiple Choice*   What is the equation of the line shown in the graph?

Ⓐ $-5x + 3y = 15$
Ⓑ $-5x + 3y = -15$
Ⓒ $5x - 3y = 15$
Ⓓ $5x + 3y = 15$
Ⓔ $5x - 3y = -15$

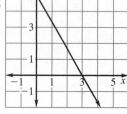

29. *Multiple Choice*   What is the y-intercept of the graph shown in Exercise 28?

Ⓐ 3   Ⓑ 5   Ⓒ -5

Ⓓ -3 Ⓔ 4

30. *Multiple Choice*   Find the slope of the line passing through the points $(7, 2)$ and $(8, 6)$.

Ⓐ -4 Ⓑ $\frac{1}{4}$ Ⓒ 4

Ⓓ $-\frac{1}{4}$ Ⓔ -6

31. *Multiple Choice*   Find the value of y so that the line passing through $(-5, 1)$ and $(3, y)$ has a slope of $-\frac{3}{8}$.

Ⓐ 2   Ⓑ $-5\frac{3}{4}$ Ⓒ $2\frac{1}{8}$

Ⓓ -2 Ⓔ $3\frac{1}{2}$

32. *Multiple Choice*   Choose the set of lines which are parallel.

Ⓐ $y = \frac{2}{3}x + 1, y = \frac{3}{2}x + 6$

Ⓑ $y = -2x + 3, y = 2x + 3$

Ⓒ $y = \frac{1}{2}x - 1, y = \frac{1}{2}(x - 2)$

Ⓓ $y = \frac{1}{5}x + 2, y = -5x - 6$

Ⓔ $y = \frac{2}{7}x, y = \frac{2}{7}x + 2$

Quantitative Comparison   In Exercises 33–36, choose the statement below that is true about the given quantity.

Ⓐ   The value of column A is greater.

Ⓑ   The value of column B is greater.

Ⓒ   The two values are equal.

Ⓓ   The relationship cannot be determined from the information given.

	Column A	Column B
33.	The constant of variation in $y = -3x$	The constant of variation in $y = \frac{1}{2}x$
34.	The slope of $y = 13x + 6$	The slope of $y = 6x + 13$
35.	The y-intercept of $y = -3x + \frac{1}{2}$	The y-intercept of $y = \frac{1}{3}x + \frac{1}{2}$
36.	$f(x) = -2x + \frac{1}{2}$ when $x = -3$	$f(x) = \frac{2}{3}x - 1$ when $x = 6$

37. *Multiple Choice*   What is the equation of the line that passes through the point $(2, -4)$ and has a slope of $\frac{1}{2}$?

Ⓐ $y = \frac{1}{2}x - 5$ Ⓑ $y = \frac{1}{2}x + 5$

Ⓒ $y = \frac{1}{2}x - 6$ Ⓓ $y = \frac{1}{2}x - 3$

Ⓔ $y = \frac{1}{2}x + 6$

38. *Multiple Choice*   What is the slope of the line perpendicular to the line $y = \frac{2}{5}x + 10$?

Ⓐ $\frac{2}{5}$ Ⓑ $-\frac{2}{5}$ Ⓒ $\frac{5}{2}$

Ⓓ $-\frac{5}{2}$ Ⓔ $-\frac{1}{10}$

39. *Multiple Choice*   Which equation in point-slope form passes through the point $(5, -2)$ and has a slope of -2?

Ⓐ $y = -2x - 12$

Ⓑ $y + 2 = -2(x - 5)$

Ⓒ $y = -2x + 8$

Ⓓ $y - 2 = -2(x - 5)$

Ⓔ $y - 2 = -2(x + 5)$

48   **Algebra 1**
Standardized Test Practice Workbook

40. *Multiple Choice* Find the standard form of the equation of the line that passes through the points $(3, 5)$ and $(-1, -2)$.

- Ⓐ $7x - 4y = 4$
- Ⓑ $7x - 4y = 1$
- Ⓒ $y = \frac{7}{4}x - \frac{1}{4}$
- Ⓓ $7x - 4y = -5$
- Ⓔ $y = \frac{7}{4}x - 5$

41. *Multi-Step Problem* The table shows the average cost of a large, one item pizza.

Year	1990	1992	1994	1996	1998
Cost	$5.60	$6.20	$6.80	$7.50	$8.30

- **a.** Draw a scatter plot of the price of a pizza in terms of the year t. Let t be the number of years since 1990.
- **b.** Write a linear model that best fits the data.
- **c.** Use the model to predict the price of a pizza in 1995. Was this linear interpolation or extrapolation?
- **d.** Use the model to predict the price in 2002.

42. *Multiple Choice* Which graph represents the solution of $-2x + 3 \geq 7$?

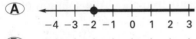

43. *Multiple Choice* Which inequality is equivalent to $x - 5 \leq 2x - 6$?

- Ⓐ $x \leq 1$
- Ⓑ $x \geq -1$
- Ⓒ $x \geq 1$
- Ⓓ $x \geq 11$
- Ⓔ $x \leq 11$

44. *Multiple Choice* Choose the solution of the inequality $6 < 2x - 3 \leq 11$.

- Ⓐ $\frac{3}{2} > x \geq 7$
- Ⓑ $\frac{9}{2} < x \leq 7$
- Ⓒ $\frac{9}{2} > x \geq 7$
- Ⓓ $\frac{3}{2} < x \leq \frac{9}{2}$
- Ⓔ $\frac{3}{2} < x \leq 7$

45. *Multiple Choice* Which numbers are solutions of the absolute-value equation $|x - 1| - 2 = 6$?

- Ⓐ 9 and -7
- Ⓑ -9 and 7
- Ⓒ 5 and -3
- Ⓓ 7 and 8
- Ⓔ -5 and -3

46. *Multiple Choice* Which point is not a solution of $y \geq 3x - 5$?

- Ⓐ $(1, 6)$
- Ⓑ $(2, 1)$
- Ⓒ $(3, 2)$
- Ⓓ $(-2, 3)$
- Ⓔ $(-1, 1)$

47. *Multiple Choice* What is the mean of 13, 5, 22, 7, 18, 11 and 12?

- Ⓐ 12.1
- Ⓑ 12.9
- Ⓒ 11.9
- Ⓓ 12.6
- Ⓔ 13.6

48. *Multiple Choice* What is the median of 13, 5, 22, 7, 18, 11 and 12?

- Ⓐ 7
- Ⓑ 11.5
- Ⓒ 13
- Ⓓ 12.5
- Ⓔ 12

49. *Multiple Choice* Find the first quartile of the collection of numbers: 6, 8, 4, 10, 9, 7, 8, 7.

- Ⓐ 6.5
- Ⓑ 7.5
- Ⓒ 8.5
- Ⓓ 6
- Ⓔ 9.5

Chapter 6

LESSON

7.1

NAME _____ DATE _____

Standardized Test Practice

For use with pages 398–403

TEST TAKING STRATEGY   Some questions involve more than one step. Reading too quickly might lead to mistaking the answer to a preliminary step for your final answer.

1. Multiple Choice   Which point represents the solution of the system of linear equations?

$y = 2x - 3$

$y = -x - 2$

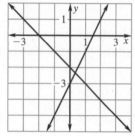

A $(0, -3)$ **B** $(0, -2)$ **C** $\left(\frac{2}{3}, -\frac{5}{3}\right)$

D $(1, 3)$ **E** $\left(\frac{1}{3}, -\frac{7}{3}\right)$

2. Multiple Choice   Which point represents the solution of the system of linear equations?

$y = -\frac{1}{2}x + 2$

$3x - 2y = 4$

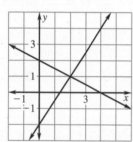

A $(1, 2)$ **B** $(2, 0)$ **C** $(2, 1)$

D $(0, -2)$ **E** $(0, 2)$

3. Multiple Choice   What is the solution of the system of linear equations?

$y = \frac{3}{2}x + 1$ $y = \frac{3}{4}x - 2$

A $(4, 5)$ **B** $(-4, -5)$

C $(-4, 5)$ **D** $(4, -5)$

E $(0, -2)$

4. Multiple Choice   What is the solution of the system of linear equations?

$y = -\frac{2}{3}x + 1$ $y = \frac{2}{3}x - 3$

A $(3, 1)$ **B** $(-3, -1)$

C $(3, -1)$ **D** $(-3, 1)$

E $(3, 0)$

5. Multiple Choice   If $y = 2x + 5$ and $4x - y = 3$, then $x =$ __?__ .

A 1   **B** 2   **C** 3

D 4   **E** 5

6. Multiple Choice   Riding the bus to and from work costs you $5.50 round trip. You are thinking about buying a used car for $2500. Gas and maintenance will cost you about $.50 per trip. How many trips to work will it take to break even?

A 510   **B** 417   **C** 450

D 500   **E** 525

7. Multiple Choice   If $y = \frac{1}{2}x - 2$ and $4x - 7y = 15$, then $x + y =$ __?__ .

A 2   **B** 1   **C** 6

D 3   **E** -1

Quantitative Comparison   In Exercises 8–10, solve the linear system. Then choose the statement below that is true about the solution of the system.

A   The value of x is greater than the value of y.

B   The value of y is greater than the value of x.

C   The values of x and y are equal.

D   The relationship cannot be determined from the given information.

8. $y = \frac{1}{2}x - 4$ $y = -\frac{3}{2}x + 1$

9. $3x + 2y = 14$ $y = -\frac{3}{2}x + 1$

10. $y = \frac{1}{2}x - 4$ $x = -1$

Chapter 7

50   **Algebra 1**
Standardized Test Practice Workbook

TEST TAKING STRATEGY **Before you give up on a question, try to eliminate some of your choices so you can make an educated guess.**

1. *Multiple Choice* Use the substitution method to determine the solution of the system of linear questions.

$x + y = 8$ $\qquad$ $y = 3x$

(A) $(1, 7)$ $\qquad$ (B) $(2, 6)$

(C) $(-2, -6)$ $\qquad$ (D) $(-1, -7)$

(E) $(3, 5)$

2. *Multiple Choice* Use the substitution method to determine the solution of the system of linear equations.

$x + 2y = -6$ $\qquad$ $2x - y = 8$

(A) $(-2, -2)$ $\qquad$ (B) $(-4, 2)$

(C) $(2, -4)$ $\qquad$ (D) $\left(-\frac{22}{5}, -\frac{26}{5}\right)$

(E) $(-10, 1)$

3. *Multiple Choice* Use the substitution method to determine the solution of the system of linear equations.

$3x + 4y = 16$ $\qquad$ $-3x + 2y = 8$

(A) $(4, 0)$ $\qquad$ (B) $(1, 4)$

(C) $(-2, 5)$ $\qquad$ (D) $(-2, 1)$

(E) $(0, 4)$

4. *Multiple Choice* Which point lies on the graph of the system?

$4x + y = 23$ $\qquad$ $x - y = 2$

(A) $(3, 5)$ $\qquad$ (B) $(5, 3)$

(C) $(4, 7)$ $\qquad$ (D) $(-2, 1)$

(E) $(0, 4)$

5. *Multiple Choice* Which point lies on the graph of the system?

$y = x - 9$ $\qquad$ $x + 2y = 0$

(A) $(6, -3)$ $\qquad$ (B) $(-3, 6)$

(C) $(5, -4)$ $\qquad$ (D) $(8, -1)$

(E) $(6, 3)$

6. *Multiple Choice* The ordered pair $(6, -8)$ is a solution of which system of equations?

(A) $3x + 2y = 4$ $\qquad$ (B) $x - 2y = 22$

$\quad$ $x - y = 14$ $\qquad$ $\quad$ $y = \frac{4}{5}x - 10$

(C) $7x + 3y = 18$ $\qquad$ (D) $2x + y = 4$

$\quad$ $x - y = -2$ $\qquad$ $\quad$ $-x - y = 2$

(E) $y = -x - 2$

$\quad$ $4x - 2y = 8$

7. *Multiple Choice* If $5x - 3y = 1$ and $x - 2y = -4$, then $xy = $ ___?___ .

(A) 5 $\qquad$ (B) 4 $\qquad$ (C) 6

(D) 8 $\qquad$ (E) 9

Quantitative Comparison In Exercises 8–10, solve the linear system. Then choose the statement below that is true about the solution of the system.

(A) The value of x is greater than the value of y.

(B) The value of y is greater than the value of x.

(C) The values of x and y are equal.

(D) The relationship cannot be determined from the given information.

8. $3x + y = 8$

$\quad$ $-4x - 3y = -9$

9. $5x + 2y = 28$

$\quad$ $y = \frac{1}{2}x + 2$

10. $3y = 2x + 1$

$\quad$ $7x - y = -13$

Chapter 7

Standardized Test Practice

For use with pages 411–417

TEST TAKING STRATEGY   Think positively during a test. This will help keep up your confidence and enable you to focus on each question.

1. *Multiple Choice*   Order the steps below to solve the system of equations using linear combinations.

   $2x + 5y = 7$ Equation 1

   $4y - 3x = 16$ Equation 2

   I.   Substitute the known variable into either of the original equations. Solve for the remaining unknown variable.

   II.   Multiply Equation 1 by 3 and Equation 2 by 2.

   III.   Multiply Equation 1 by 3 and Equation 2 by -2.

   IV.   Arrange the equations with like terms in columns.

   V.   Add the equations, combine like terms to eliminate one variable, and solve for the remaining variable.

   Ⓐ I, II, III, IV   Ⓑ IV, II, V, I

   Ⓒ II, IV, I, V   Ⓓ IV, III, V, I

   Ⓔ II, IV, I, V

2. *Multiple Choice*   Solve the system of linear equations using linear combinations.

   $5x + 3y = 26$

   $-2x + 3y = -2$

   Ⓐ $(6, -1)$ Ⓑ $(-4, -2)$ Ⓒ $(2, 5)$

   Ⓓ $(4, 2)$ Ⓔ $(7, -3)$

3. *Multiple Choice*   If $3x + 2y = 8$ and $-4x + 3y = -5$, then $x + y =$ ___?___ .

   Ⓐ 6   Ⓑ 8   Ⓒ 7

   Ⓓ 5   Ⓔ 3

4. *Multiple Choice*   If $2y - 3 = 5x$ and $y = \frac{2}{3}x + 7$, then $xy =$ ___?___ .

   Ⓐ 12   Ⓑ 25   Ⓒ 27

   Ⓓ 18   Ⓔ 14

5. *Multiple Choice*   Two hunters start from different points in the woods and hike towards their campsite. The first hunter travels a trail along the line $2y - x = 24$. The second hunter's trail follows the line $3y = -2x - 18$. They meet at the campsite. What are the coordinates of the site?

   Ⓐ $\left(-\frac{108}{7}, \frac{30}{7}\right)$ Ⓑ $\left(-\frac{30}{7}, \frac{108}{7}\right)$

   Ⓒ $(-13, 2)$ Ⓓ $(-14, 5)$

   Ⓔ $\left(-\frac{120}{7}, \frac{48}{7}\right)$

6. *Multi-Step Problem*   A fish swims 12 miles downstream in 2 hours. The fish swims back upstream to the same point, but the trip takes 3 hours. Assume the speed of the current was constant and the fish swam along the same path.

   a. Write a system of linear equations to describe the situation.

   b. Determine the speed of the fish in still water.

   c. Determine the speed of the current.

   d. *Writing*   If the speed of the current was not constant, could you still calculate the speed of the fish in still water?

NAME _____ DATE _____

Standardized Test Practice

For use with pages 418–424

TEST TAKING STRATEGY   Go back and check as much of your work as you can.

1. *Multiple Choice*   Your class is selling wrapping paper and candles for a fund raiser. Rolls of wrapping paper sell for $5 and candles sell for $4 each. A total of 526 items were sold and $2327 was raised. How many of each item was sold?

    Ⓐ   220 rolls, 306 candles

    Ⓑ   223 rolls, 303 candles

    Ⓒ   326 rolls, 200 candles

    Ⓓ   200 rolls, 326 candles

    Ⓔ   215 rolls, 311 candles

2. *Multiple Choice*   The manual for your boat engine calls for 91 octane gas. The gas stations by your house only sell 87 and 93 octane. If the boat's tank holds 30 gallons, how many gallons of each should you buy for a 91 octane mix?

    Ⓐ   15 gallons of 87 and 15 gallons of 93.

    Ⓑ   12 gallons of 87 and 18 gallons of 93.

    Ⓒ   10 gallons of 87 and 20 gallons of 93.

    Ⓓ   20 gallons of 87 and 10 gallons of 93.

    Ⓔ   18 gallons of 87 and 12 gallons of 93.

3. *Multiple Choice*   You went snow skiing for 6 hours. You skied at an average rate of 30 miles per hour and the ski lift took you back up the hill at a rate of 10 miles per hour. If your overall average rate was 23 miles per hour, how many hours did you actually spend skiing?

    Ⓐ   4   Ⓑ   3.5   Ⓒ   3.9

    Ⓓ   3.1   Ⓔ   4.5

4. *Multiple Choice*   You are taking a test worth 130 points. There are a total of 50 five-point and two-point questions. How many five-point questions are on the test?

    Ⓐ   15   Ⓑ   5   Ⓒ   20

    Ⓓ   12   Ⓔ   10

5. *Multiple Choice*   At what point do the lines $12x - 3y = 9$ and $y = -\frac{3}{2}x + 8$ intersect?

    Ⓐ $(8, 9)$ Ⓑ $(1, 4)$

    Ⓒ $\left(\frac{30}{33}, \frac{21}{33}\right)$ Ⓓ $(2, 5)$ Ⓔ $(2, 4)$

6. *Multiple Choice*   A school has two soccer seasons. There are currently 30 students playing on the spring team and participation is increasing by 2 students per year. There are currently 19 students playing on the fall team and this is increasing by 3 students per year. When will the teams be the same size?

    Ⓐ   12 years   Ⓑ   14 years

    Ⓒ   10 years   Ⓓ   11 years

    Ⓔ   8 years

Quantitative Comparison   In Exercises 7–9, solve the linear system. Then choose the statement below that is true about the solution of the system.

    Ⓐ   The value of x is greater than the value of y.

    Ⓑ   The value of y is greater than the value of x.

    Ⓒ   The values of x and y are equal.

    Ⓓ   The relationship cannot be determined from the given information.

7. $2x + 5y = -16$
   $-\frac{2}{3}x - \frac{1}{2}y = -1$

8. $3y = 10x - 14$
   $y = \frac{3}{2}x - 1$

9. $8y - 3x = 31$
   $3x + 5y = 34$

Chapter 7

NAME _____ DATE _____

Standardized Test Practice

For use with pages 426–431

TEST TAKING STRATEGY **Avoid spending too much time on one question. Skip questions that are too difficult for you and spend no more than a few minutes on each question.**

1. *Multiple Choice* How many solutions does the linear system have?

$$y = \frac{3}{2}x + 8$$

$$6x - 4y = 16$$

   Ⓐ None Ⓑ Exactly one

   Ⓒ Two Ⓓ Infinitely Many

   Ⓔ Cannot be determined

2. *Multiple Choice* How many solutions does the linear system have?

$$x = 3$$

$$y = -2$$

   Ⓐ None Ⓑ Exactly one

   Ⓒ Two Ⓓ Infinitely Many

   Ⓔ Cannot be determined

3. *Multiple Choice* How many solutions does the linear system have?

$$6x - 2y = 14$$

$$y = 3x - 9$$

   Ⓐ None Ⓑ Exactly one

   Ⓒ Two Ⓓ Infinitely Many

   Ⓔ Cannot be determined

4. *Multiple Choice* What is the solution of the system of linear equations?

$$y = \frac{1}{2}x + 8$$

$$2x + y = 18$$

   Ⓐ $(-4, -10)$ Ⓑ $(4, 10)$ Ⓒ (2.5)

   Ⓓ None Ⓔ Cannot be determined

5. *Multiple Choice* What is the solution of the system of linear equations?

$$5x - 2y = 16$$

$$y = \frac{5}{2}x - 10$$

   Ⓐ $(2, -3)$ Ⓑ $(4, 2)$

   Ⓒ $(0, -8)$ Ⓓ None

   Ⓔ Infinitely many

6. *Multiple Choice* How many solutions does the graph below have?

   Ⓐ None

   Ⓑ Exactly one

   Ⓒ Two

   Ⓓ Infinitely Many

   Ⓔ Cannot be determined

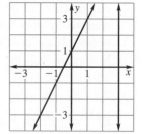

Quantitative Comparison In Exercises 7–9, solve the linear system. Then choose the statement below that is true about the solution of the system.

   Ⓐ The number of solutions in column A is greater.

   Ⓑ The number of solutions in column B is greater.

   Ⓒ The number of solutions are equal.

   Ⓓ The relationship cannot be determined from the given information.

	Column A	Column B
7.	$y = \frac{1}{2}x + 2$ $-x + 2y = 4$	$x + y = 7$ $3x + 4y = 16$
8.	$y = 3x + 6$ $y = 3x - 2$	$-3x + 2y = 2$ $y = \frac{3}{2}x + 6$
9.	$5x = 10$ $y = -7$	$7x + 2y = 12$ $14x - 24 = -4y$

NAME _____ DATE _____

Standardized Test Practice

For use with pages 432–438

TEST TAKING STRATEGY   Read all of the answer choices before deciding which is the correct one.

1. *Multiple Choice*   Choose the system of linear equalities shown in the graph.

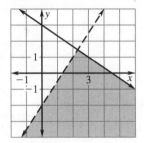

  Ⓐ $y < \frac{3}{2}x - 2$ Ⓑ $y > \frac{3}{2}x - 2$
       $y \le -\frac{2}{3}x + 3$ $y \ge -\frac{2}{3}x + 3$

  Ⓒ $y > \frac{3}{2}x - 2$ Ⓓ $y < \frac{3}{2}x - 2$
       $y \le -\frac{2}{3}x + 3$ $y \ge -\frac{2}{3}x + 3$

  Ⓔ None of these

2. *Multiple Choice*   Choose the system of linear equalities shown in the graph.

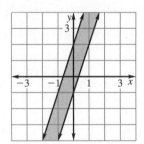

  Ⓐ $y \ge 3x + 2$ Ⓑ $y \ge 3x + 2$
       $y \le 3x - 1$ $y \le 3x - 1$

  Ⓒ $y \ge \frac{1}{3}x + 2$ Ⓓ $y \le 3x + 2$
       $y \ge \frac{1}{3}x - 1$ $y \ge 3x - 1$

  Ⓔ None of these

3. *Multiple Choice*   Which point is a solution of the system of linear inequalities?

    $y \le \frac{1}{2}x + 2$

    $y < -\frac{2}{3}x - 1$

  Ⓐ $(0, 0)$ Ⓑ $(-2, -3)$ Ⓒ $(1, 2)$

  Ⓓ $(-2, 3)$ Ⓔ $(-4, 2)$

4. *Multiple Choice*   Choose the system of linear equalities shown in the graph.

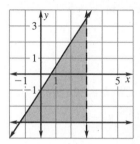

  Ⓐ $x > 3$ Ⓑ $x < 3$
       $y \ge -3$ $y \le -3$
       $y \le \frac{3}{2}x - 1$ $y \le \frac{3}{2}x - 1$

  Ⓒ $x < 3$ Ⓓ $x < 3$
       $y \le -3$ $y \ge -3$
       $y \ge \frac{3}{2}x - 1$ $y \le \frac{3}{2}x - 1$

  Ⓔ None of these

5. *Multiple Choice*   Which point is a solution of the system of linear inequalities?

    $y < x + 1$

    $y \ge 2x$

  Ⓐ $(1, -3)$ Ⓑ $(0, 3)$ Ⓒ $(2, 0)$

  Ⓓ $(-3, -8)$ Ⓔ $(-2, -2)$

6. *Multi-Step Problem*   During the summer you take two part-time jobs. The first pays $5 an hour. The second pays $8 an hour. You want to earn at least $150 a week and work 25 hours or less a week.

  a. Write a system of inequalities that model the hours you can work at each job.

  b. Graph the system.

  c. List two alternatives for dividing your time between jobs.

Algebra 1
Standardized Test Practice Workbook

Standardized Test Practice

For use with pages 450–455

TEST TAKING STRATEGY   **Be aware of how much time you have left, but keep focused on your work.**

1. *Multiple Choice*   How many factors are in the expression $5^3 \cdot 5^6$?

   A 5   **B** 9   **C** 3

   D 2   **E** 1

2. *Multiple Choice*   In the expression x^7, the 7 is called the ___?___ .

   A factor   **B** base

   C exponent   **D** product

   E term

3. *Multiple Choice*   Simplify $4^3 \cdot 4^5$. Write the answer as a power.

   A 8^8 **B** 16^8 **C** 4^{15}

   D 4^8 **E** 16^{15}

4. *Multiple Choice*   Simplify $(5^2)^7$. Write the answer as a power.

   A 25^9 **B** 25^{14} **C** 5^5

   D 5^9 **E** 5^{14}

5. *Multiple Choice*   Simplify $(-3xy^2)^4$. Write the answer as a power.

   A $81x^4y^8$ **B** $-81x^4y^8$

   C $12x^4y^8$ **D** $-12x^5y^8$

   E $81x^5y^6$

6. *Multiple Choice*   Evaluate $(a^2 \cdot a^4)^2$ when $a = 2$.

   A 256   **B** 4096   **C** 16

   D 144   **E** 128

7. *Multiple Choice*   Evaluate $-(x^3)^2(3x^2)$ when $x = -3$.

   A 19,683   **B** 6561

   C $-19,683$ **D** -6561

   E $-59,049$

8. *Multiple Choice*   The volume of a cylinder is given by $V = \pi r^2 h$, where r is the radius, h is the height and π is approximately 3.14. What is the volume of the cylinder in terms of x?

   A $94.2x^2$

   B $282.6x^2$

   C $94.2x^4$

   D $282.6x^4$

   E $887.4x^4$

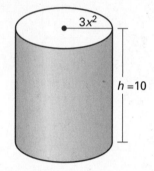

Quantitative Comparison   In Exercises 9–12, simplify each expression. Then choose the statement below that is true about the given number.

   A The number in column A is greater.

   B The number column B is greater.

   C The two numbers are equal.

   D The relationship cannot be determined with the given information.

	Column A	Column B
9.	$x^7 \cdot x^5$	$(x^7)^5$
10.	$(2x^2)^3$	$(4x^2)(2x^4)$
11.	$(3x)^2(x^3)$ when $x = -3$	$(7x^2)^3$ when $x = -2$
12.	$(ab^2)(x^3)$ when $a = 2$ and $b = -1$	$(b^2)^3(ab)$ when $a = 1$ and $b = -2$

Chapter 8

NAME _____ DATE _____

Standardized Test Practice

For use with pages 456–461

TEST TAKING STRATEGY   Learn as much as you can about the test ahead of time, such as types of questions and the topics the test will cover.

1. *Multiple Choice*   Evaluate $(5^{-3})\left(\dfrac{1}{5^{-5}}\right)(5^{-4})$.

    Ⓐ   10   Ⓑ -25 Ⓒ $\frac{1}{25}$

    Ⓓ $\frac{1}{10}$ Ⓔ $-\frac{1}{25}$

2. *Multiple Choice*   Evaluate $2^{-2} \cdot 2^{4} \cdot 2^{-3}$.

    Ⓐ $\frac{1}{4}$ Ⓑ $\frac{1}{2}$ Ⓒ   2

    Ⓓ   4   Ⓔ -2

3. *Multiple Choice*   Evaluate $(3^{3})^{-4}(3^{7})$.

    Ⓐ   243   Ⓑ $\frac{1}{243}$ Ⓒ -243

    Ⓓ   6561   Ⓔ $\frac{1}{6561}$

4. *Multiple Choice*   Rewrite $(3x^{-2}y)(6xy^{-3})$ with positive exponents.

    Ⓐ $\dfrac{18}{xy^{2}}$ Ⓑ $\dfrac{18x}{y^{2}}$ Ⓒ $\dfrac{9}{xy^{2}}$

    Ⓓ $9xy^{2}$ Ⓔ $\dfrac{18}{x^{3}y^{4}}$

5. *Multiple Choice*   Rewrite $(2x^{-2}y)(3y^{-2})$ with a positive exponent.

    Ⓐ $\dfrac{6y}{x^{2}}$ Ⓑ $\dfrac{5y}{x^{2}}$ Ⓒ $\dfrac{6y^{-3}}{x^{2}}$

    Ⓓ $\dfrac{6y^{3}}{x^{2}}$ Ⓔ $\dfrac{6}{x^{2}y}$

6. *Multiple Choice*   Evaluate $(-3)^{2}(-3)^{-2}$.

    Ⓐ   81   Ⓑ   1   Ⓒ -1

    Ⓓ   0   Ⓔ -81

7. *Multiple Choice*   Rewrite $(-3x^{3}y^{-6})(2x^{-3}y)$ with positive exponents.

    Ⓐ $\dfrac{-6x}{y^{7}}$ Ⓑ $\dfrac{-6}{xy^{5}}$ Ⓒ $\dfrac{-6x}{y^{5}}$

    Ⓓ $\dfrac{-6}{y^{5}}$ Ⓔ $\dfrac{-5}{y^{5}}$

8. *Multiple Choice*   Choose the equation of the curve shown.

    Ⓐ $y = (-1)^{x}$

    Ⓑ $y = 2^{x}$

    Ⓒ $y = 3^{x}$

    Ⓓ $y = \left(\frac{1}{2}\right)^{x}$

    Ⓔ $y = \left(\frac{1}{3}\right)^{x}$

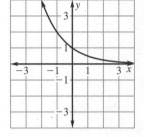

10. *Multiple Choice*   Choose the equation of the curve shown.

    Ⓐ $y = 3^{-x}$

    Ⓑ $y = 5^{-x}$

    Ⓒ $y = \left(\frac{1}{3}\right)^{-x}$

    Ⓓ $y = \left(\frac{1}{5}\right)^{-x}$

    Ⓔ $y = (-5)^{x}$

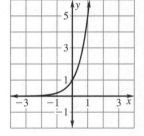

10. *Multi-Step Problem*   You started a savings account in 1996. The balance y is modeled by the equation $y = 500(1.08)^{t}$, where $t = 0$ represents the year 2000.

    a. What is the balance in 1996?

    b. What is the balance in 2000?

    c. What is the balance in 2002?

Chapter 8

Standardized Test Practice

For use with pages 463–469

TEST TAKING STRATEGY   Before you give up on a question, try to eliminate some of your choices so you can make an educated guess.

1. Multiple Choice   Evaluate $\dfrac{3^7}{3^4}$.

  Ⓐ $\dfrac{1}{27}$ Ⓑ 27   Ⓒ 9

  Ⓓ $\dfrac{1}{9}$ Ⓔ -27

2. Multiple Choice   Evaluate $\dfrac{-3^5}{(-3)^5}$.

  Ⓐ -1 Ⓑ 0   Ⓒ 3

  Ⓓ -3 Ⓔ 1

3. Multiple Choice   Evaluate $\left(-\dfrac{2}{5}\right)^2$.

  Ⓐ $-\dfrac{25}{4}$ Ⓑ $\dfrac{4}{10}$ Ⓒ $\dfrac{4}{25}$

  Ⓓ $-\dfrac{4}{25}$ Ⓔ $\dfrac{4}{5}$

4. Multiple Choice   Evaluate $\left(\dfrac{3}{4}\right)^{-3}$.

  Ⓐ $-\dfrac{9}{12}$ Ⓑ $\dfrac{12}{9}$ Ⓒ $\dfrac{27}{64}$

  Ⓓ $\dfrac{64}{27}$ Ⓔ $-\dfrac{27}{64}$

5. Multiple Choice   Simplify $\dfrac{3x^5y}{2xy^2} \cdot \dfrac{6xy^4}{4x^7}$.

  Ⓐ $\dfrac{9y^3}{4x^2}$ Ⓑ $\dfrac{9y^3}{8x^2}$ Ⓒ $\dfrac{9y^2}{4x}$

  Ⓓ $\dfrac{9x^2y^2}{8}$ Ⓔ $\dfrac{9}{4}x^2y^3$

6. Multiple Choice   Simplify $\dfrac{5xy^0}{2y^3} \cdot \dfrac{12x^2y^7}{5x}$.

  Ⓐ $6x^2y^4$ Ⓑ $6x^4y^5$ Ⓒ $\dfrac{6x^4}{y^4}$

  Ⓓ $6xy^4$ Ⓔ $6y^4$

7. Multiple Choice   Which expression simplifies to x^2?

  Ⓐ $\dfrac{x^5}{x^{-3}}$ Ⓑ $\dfrac{5x^5y}{5x^{-4}y}$ Ⓒ $\dfrac{3x^{-3}}{3x^5}$

  Ⓓ $\dfrac{x^{-7}y^2}{x^{-9}y^2}$ Ⓔ $\dfrac{x^{-9}y^2}{x^{-7}y^2}$

8. Multiple Choice   Simplify $\left(\dfrac{5x^3y}{3y^2}\right)^2 \cdot \left(\dfrac{3^2xy^3}{5x^8}\right)$.

  Ⓐ $\dfrac{5}{x}$ Ⓑ $5x$ Ⓒ $5xy$

  Ⓓ $\dfrac{5y}{x}$ Ⓔ $\dfrac{5y^2}{x}$

9. Multiple Choice   Simplify $\dfrac{(x^2y)^3}{2xy^2} \cdot \left(\dfrac{3x^4y^2}{2xy}\right)^{-2}$.

  Ⓐ $\dfrac{9x^2}{8y}$ Ⓑ $\dfrac{3}{4}x^{10}y$ Ⓒ $\dfrac{2}{9}xy$

  Ⓓ $\dfrac{2}{9}x^2y$ Ⓔ $\dfrac{2}{9xy}$

10. Multiple Choice   You roll a die three times. What is the probability that you will roll a five every time?

  Ⓐ $\left(\dfrac{1}{3}\right)^5$ Ⓑ $\left(\dfrac{1}{6}\right)^3$ Ⓒ 5^3

  Ⓓ $\left(\dfrac{1}{3}\right)^6$ Ⓔ $\left(\dfrac{1}{8}\right)^3$

Quantitative Comparison   In Exercises 11–13, simplify the expression using $x = -1$ and $y = 2$. Then choose the statement below that is true about the given number.

  Ⓐ   The number in column A is greater.

  Ⓑ   The number in column B is greater.

  Ⓒ   The two numbers are equal.

  Ⓓ   The relationship cannot be determined from the given information.

	Column A	Column B
11.	x^y	y^x
12.	$(5^x)(5yx^2)$	$\left(\dfrac{1}{2}\right)^x$
13.	$(3y^3x^{-y})(6y^x)$	$(5^yx^3) \cdot x^y$

Chapter 8

NAME _____ DATE _____

Standardized Test Practice

For use with pages 470–475

TEST TAKING STRATEGY As soon as the testing begins, start working. Keep moving and stay focused on the test.

1. *Multiple Choice* Rewrite 5.4×10^4 in decimal form.

   Ⓐ 54,000   Ⓑ 540,000

   Ⓒ 0.000054   Ⓓ 0.00054

   Ⓔ 5400

2. *Multiple Choice* Rewrite 8.2×10^{-6} in decimal form.

   Ⓐ 0.00000082   Ⓑ 0.000082

   Ⓒ 8,200,000   Ⓓ 0.0000082

   Ⓔ 820,000

3. *Multiple Choice* Rewrite 0.00000036 in scientific notation.

   Ⓐ 3.6×10^7 Ⓑ 3.6×10^{-7}

   Ⓒ 3.6×10^6 Ⓓ 3.6×10^{-6}

   Ⓔ 3.6×10^{-8}

4. *Multiple Choice* Which of the following numbers is *not* written in scientific notation?

   Ⓐ 1.2×10^8 Ⓑ 3.72×10^{-7}

   Ⓒ 8×10^5 Ⓓ 76.02×10^9

   Ⓔ 1.408×10^{-2}

5. *Multiple Choice* Evaluate the product $(7.2 \times 10^5) \cdot (6.1 \times 10^{-8})$. Write the result in scientific notation.

   Ⓐ 43.92×10^{-2} Ⓑ 4.392×10^{-3}

   Ⓒ 4.392×10^3 Ⓓ 4.392×10^2

   Ⓔ 4.392×10^{-2}

6. *Multiple Choice* Evaluate $(3.0 \times 10^{-3})^4$.

   Ⓐ 8.1×10^{-7} Ⓑ 9×10^{-12}

   Ⓒ 8.1×10^{-11} Ⓓ 8.1×10^{-12}

   Ⓔ 8.1×10^{-13}

7. *Multiple Choice* Evaluate $\dfrac{(9.6 \times 10^{-4})}{(1.2 \times 10^{-6})}$.

   Ⓐ 7×10^{10} Ⓑ 8×10^2

   Ⓒ 8.4×10^2 Ⓓ 8.4×10^{-2}

   Ⓔ 7×10^{-2}

8. *Multiple Choice* An astronomer measured the speed of a comet at 150,000 miles per hour. Write the number in scientific notation.

   Ⓐ 1.5×10^6 Ⓑ 1.5×10^{-6}

   Ⓒ 1.5×10^5 Ⓓ 1.5×10^{-5}

   Ⓔ 1.5×10^{-4}

9. *Multiple Choice* Evaluate $\dfrac{(0.002)^3}{400,000}$. Write the result in scientific notation.

   Ⓐ 2×10^{-14} Ⓑ 2×10^{-4}

   Ⓒ 1×10^{-14} Ⓓ 1×10^{-4}

   Ⓔ 2×10^{14}

Quantitative Comparison In Exercises 10–12, choose the statement below that is true about the given value.

   Ⓐ The value in column A is greater.

   Ⓑ The value column B is greater.

   Ⓒ The two values are equal.

   Ⓓ The relationship cannot be determined with the given information.

	Column A	Column B
10.	8.6×10^6	8.6×10^{-6}
11.	$(2.3 \times 10^{-5})^2$	$\dfrac{4.84 \times 10^{-4}}{8.8 \times 10^3}$
12.	$(3.2 \times 10^5)(8.6 \times 10^7)$	$\dfrac{(2 \times 10^5)^2}{(1 \times 10^{-3})}$

Chapter 8

Standardized Test Practice

For use with pages 477–482

TEST TAKING STRATEGY   Think positive during a test. This will keep up your confidence and enable you to focus on each question.

1. *Multiple Choice*   In the model
   $y = c(1 + r)^t$, $(1 + r)$ is the ___?___.

   Ⓐ time period   Ⓑ intial amount

   Ⓒ growth factor   Ⓓ growth rate

   Ⓔ percent increase

2. *Multiple Choice*   You deposit $1500 into a savings account that pays 6% interest compounded yearly. How much money is in the account after 10 years assuming you made no additional deposits or withdrawls?

   Ⓐ $1590   Ⓑ $2400

   Ⓒ $2657.34   Ⓓ $2686.27

   Ⓔ $3890.61

3. *Multiple Choice*   You deposit $750 into a savings account that pays 8% interest compounded yearly. How much money is in the account after 7 years assuming no additional deposits or withdrawls were made?

   Ⓐ $1170   Ⓑ $1285.37

   Ⓒ $1204.34   Ⓓ $1388.20

   Ⓔ $1560

4. *Multiple Choice*   Which model best represents the growth curve shown in the graph?

   Ⓐ $y = 25(1.5)^{-t}$

   Ⓑ $y = 25(1.5)^t$

   Ⓒ $y = 25(0.9)^t$

   Ⓓ $y = 50(1.1)^t$

   Ⓔ $y = 50(1.1)^{-t}$

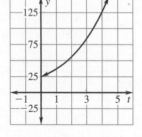

5. *Multiple Choice*   A wildlife management group releases 12 elk in a re-introduction program area. The population is expected to increase by 25% each year for the next 5 years. What is the estimated elk population after 5 years. (Round down to the nearest whole number.)

   Ⓐ 27   Ⓑ 37   Ⓒ 36

   Ⓓ 15   Ⓔ 38

6. *Multiple Choice*   You start a new job at a pay rate of $6 per hour. You expect a raise of 4% each year. After 6 raises, how much will you be earning per hour?

   Ⓐ $7.44   Ⓑ $7.02   Ⓒ $7.57

   Ⓓ $7.59   Ⓔ $8.51

7. *Multi-Step Problem*   You bought a piece of land for $50,000 in 1998 in an area of rapidly increasing population growth. You expect the land to increase in value 15% each year for the next 10 years.

   a. Write an exponential growth model for the situation.

   b. Estimate the value of the land in 2005. (Round to the nearest dollar.)

   c. Estimate the value of the land in 1996. (Round to the nearest dollar.)

   d. *Writing*   If the population growth begins to level off, what might happen to the value of the land?

Standardized Test Practice

For use with pages 484–491

TEST TAKING STRATEGY **Avoid spending too much time on one question. Skip questions that are too difficult for you, and spend no more than a few minutes on each question.**

1. *Multiple Choice*   In an exponential decay function, the decay factor is always ___?___ .

   Ⓐ   greater than zero

   Ⓑ   less than zero

   Ⓒ   greater than one

   Ⓓ   less than one

   Ⓔ   A and D

2. *Multiple Choice*   Which model is an exponential decay model?

   Ⓐ $y = 5x + 6$ Ⓑ $y = 5(1.2)^t$

   Ⓒ $y = 7(0.9)^t$ Ⓓ $y = 5 - 3t$

   Ⓔ $y = 50 - 3(1.1)^t$

3. *Multiple Choice*   Which model best represents the decay curve shown in the graph?

   Ⓐ $y = 50(0.76)^{-t}$

   Ⓑ $y = 50(0.76)^t$

   Ⓒ $y = 100(0.76)^t$

   Ⓓ $y = 100(1.32)^t$

   Ⓔ $y = 100(0.76)^{-t}$

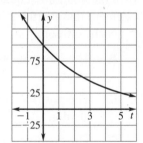

4. *Multiple Choice*   You buy a new laptop computer for $3500 in 1998. The computer depreciates at the rate of 18% per year. What is its value in 2001?

   Ⓐ $1929.79   Ⓑ $1890

   Ⓒ $1610   Ⓓ $4130

   Ⓔ $2870

5. *Multiple Choice*   You are in your state's high school tennis championship tournament. At the start of the event there are 128 participants, and each round eliminates half of the players. How many players remain after round 3?

   Ⓐ   64   Ⓑ   32   Ⓒ   16

   Ⓓ   8   Ⓔ   4

6. *Multiple Choice*   You have a large oak tree in your yard. If the tree has 10,000 leaves at the beginning of autumn, and loses 7% each day until all the leaves are off, how many leaves remain on the tree after 1 week of leaf loss?

   Ⓐ   4900   Ⓑ   9300

   Ⓒ   6957   Ⓓ   6017

   Ⓔ   5000

Quantitative Comparison   In Exercises 7–9, evaluate each function. Then choose the statement below that is true about the given values of y.

   Ⓐ   The value of y in column A is greater.

   Ⓑ   The value of y column B is greater.

   Ⓒ   The two values of y are equal.

   Ⓓ   The relationship cannot be determined with the given information.

		Column A	Column B
7.	$t = 1$	$y = 26(0.8)^{-t}$	$y = 26(1.2)^t$
8.	$t = -1$	$y = 5.6\left(\frac{2}{3}\right)^t$	$y = 5.6(1.5)^{-t}$
9.	$t = 2$	$y = 14\left(\frac{1}{2}\right)^t$	$y = 17\left(\frac{2}{5}\right)^t$

Chapter 8

NAME _____ DATE _____

Standardized Test Practice

For use with pages 503–510

TEST TAKING STRATEGY   Work as fast as you can through the easier problems, but not so fast that you are careless.

Multiple Choice   In Exercises 1–4, evaluate the expression, rounding to the nearest hundredth when necessary.

1. $\pm\sqrt{0.81}$

   A 9, −9   **B** −0.9   **C** 0.9

   D 0.9, −0.9   **E** No square roots

2. $\sqrt{-25}$

   A 5, −5   **B** 0   **C** −5

   D −12.5   **E** No square roots

3. $-\sqrt{0.3}$

   A −0.15   **B** 0   **C** −5

   D −0.55   **E** No square roots

4. $\pm\sqrt{1.25}$

   A 1.11   **B** ±1.11   **C** 1.2

   D ±1.12   **E** No square roots

5. *Multiple Choice*   Evaluate $\sqrt{b^2 - 4ac}$ when $a = 2$, $b = -3$, and $c = 4$.

   A 4.8   **B** −4.8   **C** 5.3

   D −5.3   **E** Undefined

6. *Multiple Choice*   Evaluate $\sqrt{b^2 - 4ac}$ when $a = 2$, $b = -2$, and $c = -2$.

   A 4.47   **B** −3.46   **C** 3.16

   D 2.89   **E** Undefined

7. *Multiple Choice*   Evaluate $\sqrt{b^2 - 4ac}$ when $a = 5$, $b = -6$, and $c = -3$.

   A 9.0   **B** 9.8   **C** 8.8

   D 9.4   **E** Undefined

8. *Multiple Choice*   An object is dropped from a height of 350 feet. To the nearest hundredth of a second, about how long does it take the object to hit the ground? Assume there is no air resistance.

   A 5.92 sec   **B** 18.28 sec

   C 4.68 sec   **D** 1.91 sec

   E 4.83 sec

9. *Multiple Choice*   The sales S (in dollars) of camping equipment at a small store can be modeled by $S = 52.6t^2 + 6500$, where t is the number of years since 1990. Estimate the year in which the store's sales of camping equipment will be \$12,800.

   A 2001   **B** 2005   **C** 1998

   D 1999   **E** 2003

Quantitative Comparison   In Exercises 10–13, solve the quadratic equation. Then choose the statement that is true about the positive value of x in each solution.

   A The positive value of x in column A is greater.

   B The positive value of x in column B is greater.

   C The positive values of x are equal.

   D The relationship cannot be determined from the given information.

	Column A	Column B
10.	$5x^2 - 125 = 0$	$7x^2 = 175$
11.	$2x^2 + 72 = 0$	$2x^2 = 128$
12.	$3x^2 - 25 = 2$	$2x^2 - 7 = 91$
13.	$5x^2 - 18 = 0$	$3x^2 - 17 = 23$

Algebra 1
Standardized Test Practice Workbook

TEST TAKING STRATEGY   **Some questions involve more than one step. Reading too quickly might lead to mistaking the answer to a preliminary step for your final answer.**

1. *Multiple Choice*   Which one of the following is the simplified form of $\sqrt{192}$?

   Ⓐ $3\sqrt{4}$ Ⓑ $3\sqrt{8}$ Ⓒ $4\sqrt{3}$

   Ⓓ $8\sqrt{3}$ Ⓔ $4\sqrt{12}$

2. *Multiple Choice*   Which one of the following is the simplified form of $\frac{1}{3}\sqrt{450}$?

   Ⓐ $\frac{5}{3}\sqrt{18}$ Ⓑ $\frac{2}{3}\sqrt{15}$ Ⓒ $5\sqrt{2}$

   Ⓓ $12\sqrt{2}$ Ⓔ $3\sqrt{50}$

3. *Multiple Choice*   Which one of the following is the simplified form of $-5\sqrt{20}\cdot\dfrac{\sqrt{10}}{\sqrt{30}}$?

   Ⓐ $-15\sqrt{15}$ Ⓑ $-5\sqrt{\frac{20}{3}}$

   Ⓒ $-\frac{5}{3}\sqrt{20}$ Ⓓ $-15\sqrt{60}$

   Ⓔ $\frac{-10}{3}\sqrt{15}$

4. *Multiple Choice*   Which one of the following is the simplified form of $\dfrac{\sqrt{30}\sqrt{18}}{3\sqrt{5}}$?

   Ⓐ $2\sqrt{3}$ Ⓑ $\dfrac{2\sqrt{15}}{3}$ Ⓒ 6

   Ⓓ $6\sqrt{5}$ Ⓔ $2\sqrt{5}$

5. *Multiple Choice*   Solve $7x^2 - 9 = 16$, writing the answer as a simplified radical expression.

   Ⓐ ± 3.57 Ⓑ $\pm 7\sqrt{5}$ Ⓒ $\pm\dfrac{5}{\sqrt{7}}$

   Ⓓ $\pm\dfrac{5\sqrt{7}}{7}$ Ⓔ $\pm\dfrac{7\sqrt{5}}{5}$

6. *Multiple Choice*   Find the area of the triangle using the formula $A = \frac{1}{2}bh$.

   Ⓐ $3\sqrt{15}$

   Ⓑ $15\sqrt{3}$

   Ⓒ $\frac{13}{2}\sqrt{2}$

   Ⓓ $5\sqrt{3}$

   Ⓔ $3\sqrt{5}$

   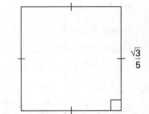

7. *Multiple Choice*   Find the area of a square whose side measures $\dfrac{\sqrt{3}}{5}$.

   Ⓐ $\frac{1}{25}\sqrt{3}$

   Ⓑ $25\sqrt{3}$

   Ⓒ $9\sqrt{5}$

   Ⓓ $\frac{3}{25}$

   Ⓔ $\frac{9}{25}$

Quantitative Comparison   In Exercises 8–11, perform the indicated operation and simplify the result. Then choose the statement below that is true about the given numbers.

   Ⓐ   The number in column A is greater.

   Ⓑ   The number in column B is greater.

   Ⓒ   The numbers are equal.

   Ⓓ   The relationship cannot be determined from the given information.

	Column A	Column B
8.	$\sqrt{6}\cdot\sqrt{-12}$	$\sqrt{\frac{72}{4}}$
9.	$4\sqrt{\frac{5}{4}}$	$18\sqrt{\frac{2}{9}}$
10.	$2\sqrt{3}\cdot\sqrt{27}$	$\dfrac{3\sqrt{3}}{\sqrt{25}}\cdot 2\sqrt{75}$
11.	$-\frac{1}{2}\sqrt{18}\cdot\frac{1}{3}\sqrt{12}$	$-3\sqrt{6}\cdot\frac{2}{3}\sqrt{18}$

TEST TAKING STRATEGY   Be aware of how much time you have left, but keep focused on your work.

1. *Multiple Choice*   The graph of $y = ax^2 + bx + c$ is a parabola whose vertex has an *x*-coordinate of _____?_____.

   Ⓐ $-\dfrac{2a}{b}$ Ⓑ $\dfrac{b}{2a}$ Ⓒ $-\dfrac{a}{2b}$

   Ⓓ $-\dfrac{b}{2a}$ Ⓔ $\dfrac{a}{2b}$

2. *Multiple Choice*   What is the *x*-coordinate of the vertex for the graph of the equation $y = -x^2 + 4x - 6$?

   Ⓐ -2 Ⓑ 2 Ⓒ $-\frac{1}{8}$
   Ⓓ $\frac{1}{8}$ Ⓔ 0

3. *Multiple Choice*   What is the *x*-coordinate of the vertex for the graph of the equation $y = \frac{1}{4}x^2 + 5x - 10$?

   Ⓐ 0 Ⓑ -10 Ⓒ 10
   Ⓓ -2 Ⓔ $-\frac{5}{4}$

4. *Multiple Choice*   What is the equation for the axis of symmetry for the equation $y = 6x^2 - 4x + 3$?

   Ⓐ $y = \frac{1}{3}$ Ⓑ $y = 3$ Ⓒ $x = \frac{1}{3}$
   Ⓓ $x = 3$ Ⓔ $x = \frac{3}{4}$

5. *Multiple Choice*   What is the *y*-coordinate of the vertex for the graph of the equation $y = -2x^2 + x - 5$?

   Ⓐ $-\frac{1}{4}$ Ⓑ $-6\frac{1}{4}$ Ⓒ $4\frac{1}{4}$
   Ⓓ $5\frac{1}{4}$ Ⓔ $-4\frac{7}{8}$

6. *Multiple Choice*   Which one of the following is the solution of $-2x^2 + 11x + 9 = y$ when $x = -2$?

   Ⓐ $y = 21$ Ⓑ $y = 22$
   Ⓒ $y = -21$ Ⓓ $y = 5$
   Ⓔ $y = -5$

7. *Multiple Choice*   Which of the following quadratic equations is represented by the graph?

   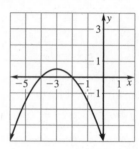

   Ⓐ $y = -2x^2 - 3x - 4$
   Ⓑ $y = \frac{1}{2}x^2 - 3x - 4$
   Ⓒ $y = 2x^2 - 3x - 4$
   Ⓓ $y = -\frac{1}{2}x^2 - 3x - 4$
   Ⓔ $y = \frac{1}{2}x^2 + 3x - 4$

Quantitative Comparison   In Exercises 8–11, find the vertex of the equation. Then choose the statement below that is true about the given values.

   Ⓐ   The value of *x* is greater.

   Ⓑ   The value of *y* is greater.

   Ⓒ   The values of *x* and *y* are equal.

   Ⓓ   The relationship cannot be determined from the given information.

8. $y = x^2 - 16$

9. $y = 3x^2 - 10x$

10. $y = x^2 - 2x + 2$

11. $y = 3x^2 + 2x + 1$

Chapter 9

LESSON
9.4

Standardized Test Practice

For use with pages 526–531

TEST TAKING STRATEGY   **Learn as much as you can about a test ahead of time, such as the types of questions and the topics that the test will cover.**

1. *Multiple Choice*   Use the graph to determine the roots of the equation.

Ⓐ   3 and -3

Ⓑ   1 and 4

Ⓒ $-\frac{1}{2}$ and 3

Ⓓ $\frac{1}{2}$ and 3

Ⓔ   4 and -3

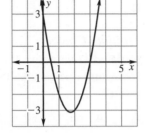

2. *Multiple Choice*   What are the x-intercepts of the graph of $y = x^2 - 3x - 18$?

Ⓐ   6 and -3 Ⓑ -6 and 3

Ⓒ   9 and 2   Ⓓ -9 and 2

Ⓔ   4 and 1

3. *Multiple Choice*   What are the x-intercepts of the graph of $y = 6x^2 - x - 1$?

Ⓐ   3 and 2   Ⓑ -3 and 2

Ⓒ $-\frac{1}{3}$ and $\frac{1}{2}$ Ⓓ $-\frac{1}{3}$ and $-\frac{1}{2}$

Ⓔ $\frac{1}{3}$ and $-\frac{1}{2}$

4. *Multiple Choice*   What are the solutions of $x^2 + x - 6 = 0$?

Ⓐ   3 and 2   Ⓑ -3 and 2

Ⓒ $\frac{1}{3}$ and 2   Ⓓ $\frac{1}{3}$ and $\frac{1}{2}$

Ⓔ -3 and 0

5. *Multiple Choice*   Which one of the following is the solution of $\frac{1}{5}x^2 = 5$?

Ⓐ   0   Ⓑ   5   Ⓒ   10

Ⓓ -20 Ⓔ   25

Quantitative Comparison   In Exercises 6–9, solve for the required information. Then choose the statement below that is true about the given information.

Ⓐ   The value of A is greater.

Ⓑ   The value of B is greater.

Ⓒ   The two values are equal.

Ⓓ   The relationship cannot be determined from the given information.

	Column A	Column B
6.	The value of x for the vertex of $x^2 - 4x - 5 = y$	The value of y for the vertex of $x^2 - 4x - 5 = y$
7.	The value of x for the vertex of $\frac{1}{2}x^2 = 8$	The sum of the roots of $\frac{1}{2}x^2 = 8$
8.	The sum of the roots of $x^2 - x - 2 = 0$	The sum of the roots of $x^2 + 2x = 15$
9.	The value of x for the vertex of $-3x^2 + 15x - 5 = 1$	The sum of the roots of $2x^2 - 5x - 3 = 0$

10. *Multiple Choice*   Choose the equation of the parabola shown in the graph below.

Ⓐ $y = 9x^2 - 1$

Ⓑ $y = 9x^2 + 1$

Ⓒ $y = \frac{1}{9}x^2 - 9$

Ⓓ $y = 9x^2 + 1$

Ⓔ $y = \frac{1}{9}x^2 - 1$

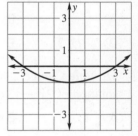

11. *Multiple Choice*   Which one of the following is a solution of $x^2 - x - 14 = 6$?

Ⓐ   0   Ⓑ   125   Ⓒ -3

Ⓓ -4 Ⓔ   4

Chapter 9

Standardized Test Practice

For use with pages 533–538

TEST TAKING STRATEGY   Go back and check as much of your work as you can.

1. *Multiple Choice*   Choose the correct form of the quadratic formula.

   (A) $x = \dfrac{b \pm \sqrt{b^2 + 4ac}}{2a}$

   (B) $x = \dfrac{-b \pm \sqrt{b^2 - 4ac}}{2a}$

   (C) $x = \dfrac{-a \pm \sqrt{a^2 - 4bc}}{2a}$

   (D) $x = \dfrac{-c \pm \sqrt{b^2 - 4ac}}{2b}$

   (E) $x = \dfrac{-b \pm \sqrt{a^2 - 4bc}}{2a}$

2. *Multiple Choice*   Choose the correct values of a, b, and c in the equation $5x^2 - x + 6 = 0$.

   (A) $a = 5$, $b = 1$, $c = 6$

   (B) $a = -5$, $b = -1$, $c = -6$

   (C) $a = 5$, $b = 0$, $c = 6$

   (D) $a = 5$, $b = -1$, $c = 0$

   (E) $a = 5$, $b = -1$, $c = 6$

3. *Multiple Choice*   What are the x-intercepts of the graph of $y = 2x^2 - x - 15$?

   (A) $\frac{5}{2}, \frac{-5}{2}$ (B) $-\frac{5}{2}, 3$ (C) $\frac{5}{3}, -3$

   (D) $-5, 3$ (E) $-5, \frac{15}{2}$

4. *Multiple Choice*   Which of the following is a solution of $2x^2 - 3x - 5 = 0$?

   (A) 1 (B) $-\frac{5}{2}$ (C) $\frac{5}{2}$

   (D) $\frac{2}{5}$ (E) 2

5. *Multiple Choice*   Which of the following is a solution of $6x^2 - 7x - 5 = 0$?

   (A) $\frac{1}{2}$ (B) $\frac{5}{3}$ (C) -2

   (D) $-\frac{3}{5}$ (E) -1

6. *Multiple Choice*   Which of the following is a solution of $7x^2 + 5x + 8 = 10$?

   (A) $\frac{2}{7}$ (B) 1 (C) $\frac{7}{2}$

   (D) $-\frac{2}{7}$ (E) $\frac{3}{7}$

7. *Multiple Choice*   You drop a rock off a bridge 30 feet above the ground into a stream. How long does it take the rock to hit the water?

   (A) 1.45 sec   (B) 1.88 sec

   (C) 1.50 sec   (D) 2.10 sec

   (E) 1.37 sec

8. *Multiple Choice*   An eagle circling a field at a height of 250 feet sees a rabbit below. The eagle dives at an initial speed of 110 feet per second. Estimate the time the rabbit has to escape.

   (A) 1.7 sec   (B) 1.8 sec

   (C) 1.6 sec   (D) 1.9 sec

   (E) 2.0 sec

9. *Multi-Step Problem*   You are on a ski-lift 50 feet high. While the lift is stopped to let people off, you accidentally knock your keys out of your pocket. They fall off the seat towards the ground.

   a. Write a vertical motion model for the path of the keys.

   b. How long will it take for the keys to hit the ground?

   c. *Critical Thinking*   What factors would change the path of the dropped keys?

Algebra 1
Standardized Test Practice Workbook

Chapter 9

TEST TAKING STRATEGY **Spend no more than a few minutes on each question.**

1. *Multiple Choice* What is the discriminant of the equation $7x^2 - 3x + 10 = 0$?

   Ⓐ 271 Ⓑ -271 Ⓒ 289

   Ⓓ -289 Ⓔ -277

2. *Multiple Choice* What is the discriminant of the equation $-3x^2 - 12x + 17 = 8$?

   Ⓐ 36 Ⓑ 348 Ⓒ -348

   Ⓓ -252 Ⓔ 252

3. *Multiple Choice* Use the discriminant to determine the number of solutions for the equation $3x^2 - 7x - 1 = 0$?

   Ⓐ 3 Ⓑ 1 Ⓒ 2

   Ⓓ Infinitely many Ⓔ None

4. *Multiple Choice* Use the discriminant to determine the number of solutions for the equation $\frac{1}{2}x^2 = 8$.

   Ⓐ 3 Ⓑ 1 Ⓒ 2

   Ⓓ Infinitely many Ⓔ None

5. *Multiple Choice* What effect does increasing the value of C by 2 have on the number of solutions for the graph of $5x^2 - 2x - 1 = 0$?

   Ⓐ Increases the number of solutions by 2.

   Ⓑ Decreases the number of solutions by 2.

   Ⓒ Increases the number of solutions by 1.

   Ⓓ Has no effect on the number of solutions.

6. *Multiple Choice* What effect does decreasing the value of C by 3 have on the number of solutions to the graph of $3x^2 + 6x + 3 = 0$?

   Ⓐ Increases the number of solutions by 2

   Ⓑ Decreases the number of solutions by 2

   Ⓒ Increases the number of solutions by 1

   Ⓓ Decreases the number of solutions by 1

   Ⓔ Has no effect on the number of solutions

7. *Multiple Choice* For the equation $2x^2 - 5x + 6 = 0$, the graph of the equation would __?__ .

   Ⓐ Have one x-intercept

   Ⓑ Have two x-intercepts

   Ⓒ Have no x-intercepts

   Ⓓ Have no y-intercepts

   Ⓔ None of these

Quantitative Comparison In Exercise 8–10, determine the number of solutions for each equation. Then choose the statement below that is true about the number of solutions.

   Ⓐ The number of solutions in column A is greater.

   Ⓑ The number of solutions in column B is greater.

   Ⓒ The number of solutions is equal.

   Ⓓ The relationship cannot be determined from the given information.

	Column A	Column B
8.	$y = 3x^2 - 2x - 6$	$y = -3x^2 + 2x - 6$
9.	$y = \frac{1}{2}x^2 - 18$	$y = -5x^2 + 10$
10.	$y = x^2 - 2x + 1$	$y = 5x^2 - 8x + 3$

Chapter 9

Algebra 1 **67**
Standardized Test Practice Workbook

Standardized Test Practice

For use with pages 548–553

TEST TAKING STRATEGY   **If you can, check your answer using a different method than you used originally to avoid making the same mistake twice.**

1. *Multiple Choice*   Which ordered pair is a solution of the inequality $y \leq 3x^2 - 5x - 6$?

   Ⓐ $(2, -3)$ Ⓑ $(1, -8)$ Ⓒ $(3, 7)$

   Ⓓ $(-2, 18)$ Ⓔ $(0, 0)$

2. *Multiple Choice*   Which ordered pair is *not* solution of the inequality $y > 8x^2 - 2x - 3$?

   Ⓐ $(-2, 35)$ Ⓑ $(0, -2)$ Ⓒ $(1, 8)$

   Ⓓ $(1, 5)$ Ⓔ $(-1, 6)$

3. *Multiple Choice*   Identify the graph of $y \geq 5x^2 + 14x - 3$.

 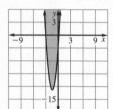 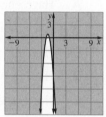

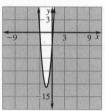 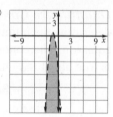

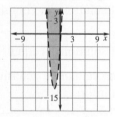

4. *Multiple Choice*   Choose the inequality which is represented by the graph.

   Ⓐ $y > 2x^2 - x - 4$

   Ⓑ $y \geq 2x^2 - x - 4$

   Ⓒ $y \leq 2x^2 - x - 4$

   Ⓓ $y < 2x^2 - x - 4$

   Ⓔ $y < -2x^2 - x - 4$

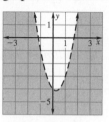

5. *Multiple Choice*   A volcano is catapulting rocks from its crator. The rocks traveling the furthest are following a path along $y = -x^2 + 60x - 80$, where x and y are measured in feet. To be safe from the falling rocks, you should be standing in the area defined by which of the inequalities?

   Ⓐ $y > -x^2 + 60x - 80$

   Ⓑ $y \geq -x^2 + 60x - 80$

   Ⓒ $y < -x^2 + 60x - 80$

   Ⓓ $y \leq -x^2 + 60x - 80$

   Ⓔ   None of these

6. *Multi-Step Problem*   A backhoe is digging a trench. On its first sweep, the shovel follows a path modeled by the equation $y = x^2 - 2x - 3$ where x is the length in feet and y is the depth in feet of the hole.

   a. What is the deepest point of the hole?

   b. What is the length of the hole?

   c. *Critical Thinking*   A water line runs perpendicular to the ditch the backhoe is digging. If the line is buried $3\frac{1}{2}$ feet down, did the backhoe hit it on the first sweep of the shovel? Explain.

Chapter 9

Standardized Test Practice

For use with pages 554–560

TEST TAKING STRATEGY   **Some questions involve more than one step. Reading too quickly might lead to mistaking the answer to a preliminary step for your final answer.**

Multiple Choice   For Exercises 1–6, name the type of model suggested by the graph or data collection using the following.

- Ⓐ Quadratic
- Ⓑ Exponential Growth
- Ⓒ Exponential Decay
- Ⓓ Linear
- Ⓔ None of these

1. 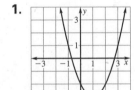 **2.**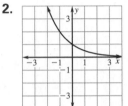

3.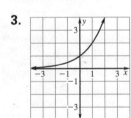

4. $(-2, 2), (-1, 2.5), (0, 3), (1, 3.5), (2, 4.3),$ $(3, 5.2)$

5. $(-1, 4), (0, 6), (1, 8), (2, 10), (3, 12), (4, 14)$

6. $(-5, 3), (-3, 1), (-1, 2), (1, -2), (3, 4),$ $(4, 1)$

7. *Multiple Choice*   You bought a car for $12,000 in 1996. It has decreased in value by 14.5 % each year since then. Which model best fits the situation, where t is the number of years since 1996?

- Ⓐ $y = 0.145t + 12,000$
- Ⓑ $y = 12,000 - 0.145t$
- Ⓒ $y = 12,000(1.45)^t$
- Ⓓ $y = 12,000(0.855)^t$
- Ⓔ $y = 0.145t^2 + 12,0000$

8. *Multiple Choice*   Name the model that the graph of $y = \frac{1}{5}x^2 + 10$ represents.

- Ⓐ Quadratic
- Ⓑ Exponential Growth
- Ⓒ Exponential Decay
- Ⓓ Linear
- Ⓔ None of these

9. *Multiple Choice*   Name the model that the graph of $y = 2.6(0.7)^t$ represents.

- Ⓐ Quadratic
- Ⓑ Exponential Growth
- Ⓒ Exponential Decay
- Ⓓ Linear
- Ⓔ None of these

10. *Multi-Step Problem*   Use the collection of data below.

x	-4	-3	-1	0	1	2
y	0	-7	-15	-16	-15	-12

   a. Make a graph of the data.

   b. Name the type of model that best fits the data.

   c. Write a model that best fits the data.

Chapter 9

NAME _____ DATE _____

Standardized Test Practice

For use with pages 576–582

TEST TAKING STRATEGY **Read all of the answer choices before deciding which is the correct one.**

1. *Multiple Choice* Classify the equation $7 + 5x + 2x^3 + 3x^2$ by degree and by the number of terms.

- Ⓐ Quadratic polynomial
- Ⓑ Quadratic binomial
- Ⓒ Cubic binomial
- Ⓓ Cubic trinomial
- Ⓔ Cubic polynomial

2. *Multiple Choice* Which of the following is equal to $(7x^2 + 3) + (5x^2 + 8)$?

- Ⓐ $12x^2 + 11$
- Ⓑ $12x^2 + 5$
- Ⓒ $12x^2$
- Ⓓ $12x^4 - 5$
- Ⓔ $12x^4 + 11$

3. *Multiple Choice* Which of the following is equal to $(3x^2 + 7x - 6) + (5x^3 + 3x^2 + 10x - 8)$?

- Ⓐ $8x^3 + 10x^2 + 10x - 14$
- Ⓑ $8x^3 + 10x^2 + 10x + 2$
- Ⓒ $8x^3 + 3x^2 + 17x - 14$
- Ⓓ $5x^3 + 6x^2 + 17x - 14$
- Ⓔ $5x^3 + 6x^2 + 17x + 2$

4. *Multiple Choice* Which of the following is equal to $(4x^2 + 7x - 6) - (3x^2 + 9x - 8)$?

- Ⓐ $x^2 + 2x + 2$
- Ⓑ $x^2 - 2x + 2$
- Ⓒ $x^2 - 2x - 14$
- Ⓓ $7x^2 + 18x - 14$
- Ⓔ $7x^2 + 18x + 2$

5. *Multiple Choice* Which of the following is equal to $(5x^3 + 2x^2 - 7x + 9) - (-4x^3 + 3x^2 + 2x + 1)$?

- Ⓐ $x^3 - x^2 - 5x - 8$
- Ⓑ $9x^3 - x^2 - 9x + 8$
- Ⓒ $x^3 + 5x^2 - 9x + 1$
- Ⓓ $9x^3 + 5x^2 - 9x + 1$
- Ⓔ $x^3 - x^2 - 9x + 8$

6. *Multiple Choice* You want to fence in a pool and deck area. Find the expression which correctly determines the amount of fencing required to complete the job.

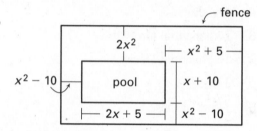

- Ⓐ $5x^2 + 3x$
- Ⓑ $10x^2 + 6x$
- Ⓒ $10x^2 + 3x + 10$
- Ⓓ $5x^2 + 6x + 10$
- Ⓔ $10x^2 + 8x + 20$

7. *Multi-Step Problem* A company is keeping track of 3 different divisions. The sales S in thousands of dollars for the first two divisions and the whole company are modeled below, where t is the number of years since 1998.

Division 1: $S = 12.2t^2 + 6.3t + 65$

Division 2: $S = 13.5t^2 - 4.3t + 75$

Company: $S = 34.6t^2 + 3.2t + 213$

a. Find a model that represents division 3.

b. Predict sales for division 3 in 2000.

c. *Critical Thinking* Sales are $350,000 one year, but division 2 did not contribute. Write an equation in terms of t to model this situation.

LESSON

10.2

NAME _____ DATE _____

Standardized Test Practice

For use with pages 584–589

TEST TAKING STRATEGY   **Before you give up on a question, try to eliminate some of your choices so you can make an educated guess.**

1. *Multiple Choice*   Which of the following is equal to $3x^2(5x - x^2 + 7)$?

   (A) $8x^3 - 3x^4 + 21x^2$

   (B) $-3x^4 + 8x^2 + 21x^2$

   (C) $-3x^4 + 8x^3 + 10x^2$

   (D) $-3x^4 + 15x^3 + 21x^2$

   (E) $3x^4 + 15x^3 + 21x^2$

2. *Multiple Choice*   Which of the following is equal to $-2y^3(2y^2 - 5y + 6)$?

   (A) $4y^2 - 10y^4 + 12y^3$

   (B) $-4y^6 + 10y^4 - 12y^3$

   (C) $-4y^6 - 10y^4 - 6y^3$

   (D) $y^5 - 7y^4 + 4y^3$

   (E) $-4y^5 + 10y^4 - 12y^3$

3. *Multiple Choice*   Which of the following is equal to $(b - 8)(b + 2)$?

   (A) $b^2 - 6b - 16$ (B) $b^2 + 6b - 16$

   (C) $b^2 - 6b - 6$ (D) $b^2 - 6b + 6$

   (E) $b^2 + 10b - 16$

4. *Multiple Choice*   Which of the following is equal to $(x - 14)(x - 3)$?

   (A) $x^2 - 17 - 17$ (B) $x^2 - 11x + 42$

   (C) $x^2 - 17x - 42$ (D) $x^2 - 11x - 42$

   (E) $x^2 - 17x + 42$

5. *Multiple Choice*   Find the product of $(x + 5)(3x^2 - 2x + 1)$.

   (A) $3x^3 + 13x^2 - 10x + 5$

   (B) $3x^3 + 13x^2 - 10x + 6$

   (C) $3x^3 + 6x^2 - 6x + 6$

   (D) $3x^3 + 13x^2 - 9x + 5$

   (E) $3x^3 + 17x^2 - 9x + 5$

6. *Multiple Choice*   Find the product of $(4x^2 + 6x - 7)(2x - 3)$.

   (A) $8x^3 - 24x^2 - 32x + 21$

   (B) $8x^3 - 12x^2 - 20x + 21$

   (C) $8x^3 - 32x + 21$

   (D) $8x^3 - 4x - 21$

   (E) $8x^3 + 24x^2 - 32x + 21$

7. *Multiple Choice*   Find the area of the parallelogram by using the formula $A = bh$.

   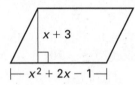

   (A) $x^3 + 5x^2 + 5x - 3$

   (B) $x^3 + 6x^2 - 6x + 3$

   (C) $x^3 + 2x^2 - x + 3$

   (D) $x^2 + 3x + 2$

   (E) $x^2 + 3x - 2$

Quantitative Comparison   In Exercises 8–10, evaluate the expression for $x = 2$ and $y = -3$. Then choose the statement below that is true about the results.

   (A) The number in column A is greater.

   (B) The number in column B is greater.

   (C) The two numbers are equal.

   (D) The relationship cannot be determined from the given information.

	Column A	Column B
8.	$2x(x^2 + 3x + 1)$	$-2y(3y^2 + 2y)$
9.	$(x + 1)(x - 6)$	$(3y + 2)(2y + 5)$
10.	$(2x - 9)(x - 5)$	$(y + 3)(y^2 + 2y + 5)$

NAME _____ DATE _____

Standardized Test Practice

For use with pages 590–596

TEST TAKING STRATEGY   **Go back and check as much of your work as you can.**

1. *Multiple Choice*   Which of the following is equal to $(x + 3)^2$?

Ⓐ $x^2 + 9$ Ⓑ $x^2 - 6x + 9$

Ⓒ $x^2 - 9$ Ⓓ $x^2 - 6x - 9$

Ⓔ $x^2 + 6x + 9$

2. *Multiple Choice*   Which of the following is equal to $(2x - 3)^2$?

Ⓐ $4x^2 + 9$ Ⓑ $4x^2 - 12x + 9$

Ⓒ $4x^2 - 9$ Ⓓ $4x^2 - 12x - 9$

Ⓔ $4x^2 - 10x + 9$

3. *Multiple Choice*   Which of the following is equal to the expression $25x^2 - 70x + 49$?

Ⓐ $(7x + 5)^2$ Ⓑ $(5x + 7)^2$

Ⓒ $(7x - 5)^2$ Ⓓ $(5x - 7)^2$

Ⓔ $(5x - 7)(5x + 7)$

4. *Multiple Choice*   Find the area of the circle below. (*Hint:* $A = \pi r^2$.)

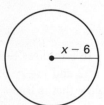

Ⓐ $A = \pi(x^2 + 6)$

Ⓑ $A = \pi(x^2 + 12x + 12)$

Ⓒ $A = \pi(x^2 - 12x - 12)$

Ⓓ $A = \pi(x^2 + 12x + 36)$

Ⓔ $A = \pi(x^2 - 12x + 36)$

5. *Multiple Choice*   Which of the following is equal to $(5x + 3)(5x - 3) - (2x - 5)^2$?

Ⓐ $21x^2 - 20x + 16$ Ⓑ $29x^2 - 20x + 16$

Ⓒ $21x^2 - 20x - 34$ Ⓓ $21x^2 - 34$

Ⓔ $21x^2 + 20x - 34$

6. *Multiple Choice*   Which of the following represents the area of the figure shown?

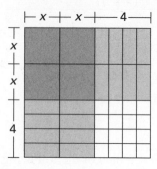

Ⓐ $4x^2 - 16$

Ⓑ $4x^2 + 16x + 16$

Ⓒ $4x^2 - 16x + 16$

Ⓓ $4x^2 + 8x - 16$

Ⓔ $4x + 8$

Quantitative Comparison   In Exercises 7–9, evaluate the expression for the given values. Then choose the statement below that is true about the results.

Ⓐ   The number in column A is greater.

Ⓑ   The number in column B is greater.

Ⓒ   The two numbers are equal.

Ⓓ   The relationship cannot be determined from the given information.

	Column A	Column B
7.	$(x + 2)^2$ when $x = 3$	$(y - 2)^2$ when $y = 3$
8.	$(a - b)^2$ when $a = 8$ and $b = -4$	$(a + b)^2$ when $a = 8$ and $b = -4$
9.	$(a - b)(a + b)$ when $a = 5$ and $b = -4$	$(a^2 - b^2)^2$ when $a = 5$ and $b = -4$

LESSON 10.4

Standardized Test Practice

For use with pages 597–602

TEST TAKING STRATEGY   **Some questions involve more than one step. Reading too quickly might lead to mistaking the answer to a preliminary step for your final answer.**

1. *Multiple Choice*   Which of the following is one of the solutions of the equation $(x - 3)(x + 8) = 0$?

   (A) 8   (B) -3 (C) 0

   (D) $\frac{3}{8}$ (E) 3

2. *Multiple Choice*   Choose the solutions of the equation $(4x - 3)(x + 1) = 0$.

   (A) $1, -\frac{3}{4}$ (B) $-1, \frac{3}{4}$ (C) $1, -\frac{4}{3}$

   (D) $1, \frac{4}{3}$ (E) $-1, \frac{4}{3}$

3. *Multiple Choice*   Which of the following are solutions of the equation $(2x - 5)^2 = 0$?

   (A) $\frac{5}{2}$ (B) $\frac{5}{2}, \frac{-5}{2}$ (C) $\frac{2}{5}$

   (D) $\frac{2}{5}, -\frac{2}{5}$ (E) 5

4. *Multiple Choice*   Choose the solutions of the equation $(3x - 1)(x + 6)(2x + 7) = 0$.

   (A) $\frac{1}{3}, 6, \frac{7}{2}$ (B) $3, -6, -7$

   (C) $\frac{1}{3}, -6, \frac{-7}{2}$ (D) $-\frac{1}{3}, 6, \frac{7}{2}$

   (E) $3, -6, -\frac{2}{7}$

5. *Multiple Choice*   Choose the equation whose graph is shown.

   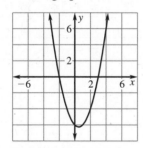

   (A) $y = (x + 2)(x + 3)$

   (B) $y = (x - 2)(x + 3)$

   (C) $y = (x - 3)(x + 2)$

   (D) $y = (x - 3)^2$

   (E) $y = (x - 3)(x - 2)$

6. *Multiple Choice*   What are the coordinates of the vertex of the graph of $y = (x - 4)(2x + 1)$?

   (A) $\left(-4, \frac{1}{2}\right)$ (B) $\left(\frac{5}{3}, -8\frac{1}{3}\right)$

   (C) $\left(-\frac{4}{3}, -8\frac{1}{3}\right)$ (D) $\left(\frac{7}{4}, -10\frac{1}{8}\right)$

   (E) $\left(4, -\frac{1}{2}\right)$

7. *Multiple Choice*   What are the coordinates of the vertex of the graph of $y = (-2x + 1)(x - 3)$?

   (A) $(0, 21)$ (B) $(7, -3)$

   (C) $\left(\frac{7}{4}, 9\frac{1}{8}\right)$ (D) $(3, -7)$

   (E) $(2, 25)$

8. *Multi-Step Problem*   A dome tent's arch is modeled by $y = -0.18(x - 6)(x + 6)$, where x and y are measured in feet.

   a. How wide is the tent at the base?

   b. How high is the tent at its highest point?

   c. *Critical Thinking*   The base of the tent is a square. Each person requires 18 square feet to sleep comfortably. If you require an area of 25 square feet to store gear and supplies, how many people can sleep in the tent?

NAME _____ DATE _____

Standardized Test Practice

For use with pages 604–609

TEST TAKING STRATEGY   Be aware of how much time you have left, but keep focused on your work.

1. *Multiple Choice*   Which of the following is a correct factorization of $x^2 + 4x - 12$?

   (A) $(x + 3)(x - 4)$ (B) $(x + 6)(x - 2)$

   (C) $(x - 6)(x + 2)$ (D) $(x - 3)(x + 4)$

   (E) Cannot be factored

2. *Multiple Choice*   Which of the following is a correct factorization of $x^2 - 5x + 6$?

   (A) $(x - 6)(x - 1)$ (B) $(x + 6)(x - 1)$

   (C) $(x - 2)(x - 3)$ (D) $(x + 2)(x + 3)$

   (E) Cannot be factored

3. *Multiple Choice*   Which of the following is a correct factorization of $x^2 + 10 - 9$?

   (A) $(x - 9)(x + 1)$ (B) $(x + 9)(x - 1)$

   (C) $(x - 9)(x - 1)$ (D) $(x - 3)(x - 3)$

   (E) Cannot be factored

4. *Multiple Choice*   Use factoring to solve the equation $x^2 + 4x = 21$.

   (A) $-7, -3$ (B) $7, 3$ (C) $-7, 3$

   (D) $7, -3$ (E) Cannot be factored

5. *Multiple Choice*   Use factoring to solve the equation $x^2 + 2x - 20 = 28$.

   (A) $6, -8$ (B) $6, 8$

   (C) $4, -12$ (D) $-4, 12$

   (E) Cannot be factored

6. *Multiple Choice*   Use factoring to solve the equation $x^2 + 13x + 12 = -24$.

   (A) $9, -4$ (B) $9, 4$

   (C) $-9, 4$ (D) $-9, -4$

   (E) Cannot be factored

7. *Multiple Choice*   The area of a rectangle is given by $x^2 - 4x - 21$. Use factoring to find expressions for the dimensions of the rectangle.

   (A) $x - 7, x - 3$ (B) $x + 7, x - 3$

   (C) $x - 21, x + 1$ (D) $x - 7, x + 3$

   (E) Cannot be determined

8. *Multiple Choice*   The area of a rectangle is 48. If the length is 8 less than its width, what is its width?

   (A) 6   (B) 4   (C) 12

   (D) 8   (E) 9

Quantitative Comparison   In Exercises 9–12, solve the equation by factoring and add the solutions. Then choose the statement that is true about the values.

   (A) The value in column A is greater.

   (B) The value in column B is greater.

   (C) The two values are equal.

   (D) The relationship cannot be determined from the given information.

	Column A	Column B
9.	$x^2 - 3x - 14 = 14$	$x^2 + 5x + 6 = 0$
10.	$x^2 - 2x - 20 = 28$	$x^2 + 5x - 36 = 0$
11.	$x^2 - 4x + 10 = -8$	$x^2 - 4x - 22 = 0$
12.	$x^2 - 4x - 12 = 0$	$x^2 = -25$

Algebra 1
Standardized Test Practice Workbook

LESSON 10.6

Standardized Test Practice

For use with pages 611–617

TEST TAKING STRATEGY As soon as the test begins, start working. Keep moving and stay focused on the test.

1. ***Multiple Choice*** Which of the following is a correct factorization of $10x^2 + 19x + 6$?

  Ⓐ $(2x + 2)(5x + 3)$

  Ⓑ $(5x + 2)(2x + 3)$

  Ⓒ $(5x - 3)(2x + 2)$

  Ⓓ $(10x + 6)(x + 1)$

  Ⓔ $(10x + 1)(x + 6)$

2. ***Multiple Choice*** Which of the following is a correct factorization of $21x^2 + 8x - 4$?

  Ⓐ $(21x - 4)(x + 1)$

  Ⓑ $(7x - 2)(3x - 2)$

  Ⓒ $(7x + 2)(3x - 2)$

  Ⓓ $(7x - 2)(3x + 2)$

  Ⓔ $(7x - 4)(3x + 1)$

3. ***Multiple Choice*** Which of the following is a correct factorization of $27x^2 - 69x + 40$?

  Ⓐ $(3x - 5)(9x + 8)$

  Ⓑ $(9x - 5)(3x - 8)$

  Ⓒ $(3x - 5)(9x - 8)$

  Ⓓ $(3x - 4)(9x - 10)$

  Ⓔ $(9x - 4)(3x - 10)$

4. ***Multiple Choice*** Which of the following is a solution of the equation $14x^2 - 23x - 30 = 0$?

  Ⓐ $-\frac{5}{2}$ Ⓑ $\frac{5}{2}$ Ⓒ $\frac{6}{7}$

  Ⓓ $-\frac{2}{7}$ Ⓔ $\frac{15}{2}$

5. ***Multiple Choice*** Which of the following is a solution of the equation $24x^2 + 26x = 5$?

  Ⓐ $\frac{1}{6}$ Ⓑ $\frac{4}{5}$ Ⓒ $\frac{5}{4}$

  Ⓓ $\frac{5}{6}$ Ⓔ $-\frac{1}{4}$

6. ***Multiple Choice*** Which one of the following equations cannot be solved by factoring with integer coefficients?

  Ⓐ $6x^2 + 5x - 6 = 0$

  Ⓑ $16x^2 - 26x = -3$

  Ⓒ $10x^2 + 9x + 3 = 0$

  Ⓓ $15x^2 = 12 - 8x$

  Ⓔ $6x^2 + 14 = 31x$

7. ***Multiple Choice*** Multiply each side of $0.33x^2 + 0.71x - 0.14 = 0$ an appropriate power of ten to obtain integer coefficients. Then solve the equation by factoring.

  Ⓐ $\frac{7}{11}, \frac{2}{3}$ Ⓑ $\frac{7}{11}, -\frac{2}{3}$

  Ⓒ $12\frac{7}{11}, \frac{2}{3}$ Ⓓ $-\frac{2}{11}, \frac{7}{3}$

  Ⓔ $\frac{2}{11}, -\frac{7}{3}$

8. ***Multiple Choice*** The area of a circle is given by $A = \pi(4x^2 + 28x + 49)$. The radius of the circle is ___?___ .

  Ⓐ $2x - 7$ Ⓑ $7x - 2$

  Ⓒ $7x + 2$ Ⓓ $2x + 7$

  Ⓔ $-2x + 7$

9. ***Multi-Step Problem*** A diver jumps off a 20 foot high diving board with an initial upward velocity of 4 feet per second. Use the vertical motion model $n = -16t^2 + vt + s$.

  a. Use the information above to model the vertical motion of the diver.

  b. Find the solutions by factoring the model you found in part (a).

  c. ***Critical Thinking*** Explain why only one of the solutions is reasonable.

NAME _____ DATE _____

Standardized Test Practice

For use with pages 619–624

TEST TAKING STRATEGY   **Think positively during a test. This will help keep up your confidence and enable you to focus on each question.**

1. *Multiple Choice*   Which of the following is a factorization of $x^2 - 16x + 64$?

   Ⓐ $(x + 8)(x - 8)$ Ⓑ $(x + 8)^2$

   Ⓒ $2(x - 4)(x - 4)$ Ⓓ $(x - 8)^2$

   Ⓔ $2(x + 4)(x - 4)$

2. *Multiple Choice*   Which of the following is a factorization of $25x^2 - 16$?

   Ⓐ $(5x - 4)(5x + 4)$ Ⓑ $(5x - 4)^2$

   Ⓒ $(5x - 16)(5x - 1)$ Ⓓ $(5x + 4)^2$

   Ⓔ $(5x + 16)(5x + 1)$

3. *Multiple Choice*   Which of the following is a factorization of $36x^2 + 84x + 49$?

   Ⓐ $(6x - 7)(6x + 7)$ Ⓑ $(6x + 7)^2$

   Ⓒ $2(9x + 3)(2x + 4)$ Ⓓ $(6x - 7)^2$

   Ⓔ   Cannot be factored

4. *Multiple Choice*   Which of the following is a factorization of $20x^2 - 125$?

   Ⓐ $5(2x - 10)(2x + 10)$

   Ⓑ $5(2x - 5)^2$

   Ⓒ $5(2x - 5)(2x + 5)$

   Ⓓ $5(2x + 5)^2$

   Ⓔ   Cannot be factored

5. *Multiple Choice*   Which of the following is a factorization of $-75x^2 - 30x - 3$?

   Ⓐ $-3(5x + 1)(5x - 1)$ Ⓑ $-3(5x - 1)^2$

   Ⓒ $(15x - 3)(-5x - 1)$

   Ⓓ $-3(5x + 1)^2$

   Ⓔ   Cannot be factored

6. *Multiple Choice*   Which of the following is a solution of $2x^2 - 98 = 0$?

   Ⓐ 0   Ⓑ 14   Ⓒ 12

   Ⓓ 7   Ⓔ   Cannot be factored

7. *Multiple Choice*   Which of the following is a solution of $4x^2 + \frac{4}{3}x + \frac{1}{9} = 0$?

   Ⓐ $-\frac{1}{6}$ Ⓑ $\frac{1}{6}$ Ⓒ $-\frac{2}{3}$

   Ⓓ $\frac{2}{3}$ Ⓔ $\frac{3}{2}$

8. *Multiple Choice*   An object is propelled from the ground with an initial upward velocity of 80 feet per second. How long does it take to reach a height of 100 feet?

   Ⓐ 1.9 sec   Ⓑ 2.6 sec

   Ⓒ 1.8 sec   Ⓓ 2.1 sec

   Ⓔ 2.5 sec

9. *Multiple Choice*   Determine the diameter D necessary for a wire rope to lift a 10.24 ton load safely.

   Ⓐ 4 inches   Ⓑ 1.6 inches

   Ⓒ 1.8 inches   Ⓓ 2 inches

   Ⓔ 6.4 inches

Quantitative Comparison   In Exercises 10–11, solve the equation by factoring and add the solutions. Then choose the statement that is true about the values.

   Ⓐ   The value in column A is greater.

   Ⓑ   The value in column B is greater.

   Ⓒ   The two values are equal.

   Ⓓ   The relationship cannot be determined from the given information.

	Column A	Column B
10.	The solution of $x^2 - 16 = 0$	The solution of $9x^2 - 25 = 0$
11.	The solution of $36x^2 - 16 = 0$	The solution of $x^2 - 10x + 25 = 0$

LESSON 10.8

NAME _____ DATE _____

Standardized Test Practice

For use with pages 625–632

TEST TAKING STRATEGY Work as fast as you can through the easier problems, but not so fast that you are careless.

1. Multiple Choice Which expression shows the greatest common factor removed from $24x^4 - 48x^3 + 60x^2$?

Ⓐ $12(2x^4 - 4x^3 - 5x^2)$

Ⓑ $12x^2(2x^2 - 4x - 5)$

Ⓒ $12x(2x^3 - 4x^2 - 5x)$

Ⓓ $x^2(24x^2 - 48x + 60)$

Ⓔ $6x^2(8x^2 - 8x - 10)$

2. Multiple Choice Which of the following is the complete factorization of $21y^4 + 49y^3 + 14y^2$?

Ⓐ $y^2(7y + 14)(3y + 1)$

Ⓑ $7y(y + 2)(3y^2 + y)$

Ⓒ $7y^2(y + 2)(3y + 1)$

Ⓓ $y^2(7y + 2)(3y + 7)$

Ⓔ $7y^2(y + 1)(3y + 2)$

Quantitative Comparison In Exercises 3–5, solve the equation by factoring and add the solutions. Choose the statement that is true about the given value.

Ⓐ The value in column A is greater.

Ⓑ The value in column B is greater.

Ⓒ The two values are equal.

Ⓓ The relationship cannot be determined from the given information.

6. Multiple Choice Which of the following is the complete factorization of $x^3 + 5x + 3x^2 + 15$?

Ⓐ $(x + 3)(x - 2)(x + 3)$

Ⓑ $(x + 3)(x + 5)(x - 1)$

Ⓒ $(x - 3)(x^2 + 5)$

Ⓓ $(x + 3)(x^2 + 5)$

Ⓔ $(x + 5)(x^2 + 3)$

7. Multiple Choice The length of a box is 5 inches less than twice the width. The height is 4 inches more than three times the width. The box has a volume of 520 cubic inches. Which of the following equations can be used to find the height of the box?

Ⓐ $w(2l - 5)(3h + 4) = 520$

Ⓑ $w(2w - 5)(3w + 4) = 520$

Ⓒ $w(2w + 5)(3w - 4) = 520$

Ⓓ $w(2w - 5)(3w - 4) = 520$

Ⓔ $l(5 - 2w)(3h + 4) = 520$

8. Multiple Choice Which equation cannot be solved using the factoring method?

Ⓐ $4x^3 + 4x^2 - x - 1 = 0$

Ⓑ $x^3 + 2x^2 + 3x + 6 = 0$

Ⓒ $90x^2 + 15x - 30 = 0$

Ⓓ $4x^3 - 4x^2 - 24x = 0$

Ⓔ $6x^3 - 15x^2 - 39x = 0$

	Column A	Column B
3.	$5x^2 + 15x + 10 = 0$	$2x^3 - 4x^2 - 6x = 0$
4.	$2x^2 + 2x + 5x + 5 = 0$	$x^3 + 5x^2 - 9x - 45 = 0$
5.	$9x^3 - 9x^2 - 4x + 4 = 0$	$5x^3 - 5x^2 - 30x = 0$

NAME _____ DATE _____

Standardized Test Practice

For use with pages 643–648

TEST TAKING STRATEGY Some questions involve more than one step. Reading too quickly might lead to mistaking the answer to a preliminary step for your final answer.

1. Multiple Choice What are the means of the proportion $\frac{5}{8} = \frac{10}{16}$?

Ⓐ 5 and 8 Ⓑ 5 and 16

Ⓒ 5 and 10 Ⓓ 8 and 16

Ⓔ 8 and 10

2. Multiple Choice Which of the following is the solution of $\frac{5}{16} = \frac{x}{12}$?

Ⓐ $6\frac{2}{3}$ Ⓑ $\frac{4}{15}$ Ⓒ $3\frac{3}{4}$

Ⓓ $3\frac{1}{2}$ Ⓔ $1\frac{1}{16}$

3. Multiple Choice Which of the following is the solution of $\frac{4}{3x} = \frac{14}{21}$?

Ⓐ 2 Ⓑ $\frac{1}{2}$ Ⓒ 8

Ⓓ $\frac{1}{8}$ Ⓔ 9

4. Multiple Choice Which of the following is the solution of $\frac{2x}{5} = \frac{40}{x}$?

Ⓐ 25 Ⓑ 50 Ⓒ 4.5

Ⓓ 10 Ⓔ 20

5. Multiple Choice Which of the following is the solution of $\frac{x-2}{10} = \frac{x}{15}$?

Ⓐ $\frac{2}{5}$ Ⓑ 6 Ⓒ −6

Ⓓ $-\frac{2}{5}$ Ⓔ 7

6. Multiple Choice Which of the following is a solution of $\frac{-3}{y-3} = \frac{y}{-6}$?

Ⓐ 3 Ⓑ −6 Ⓒ 6

Ⓓ 2 Ⓔ 4

7. Multiple Choice You are looking at a map to see how much further it is to your destination. The distance on the map is 1.6 inches. If the map has a scale of 1 mile is $\frac{1}{32}$ of an inch, how much further do you have to go?

Ⓐ 51.2 miles Ⓑ 15.6 miles

Ⓒ 23.4 miles Ⓓ 20.0 miles

Ⓔ 57.5 miles

8. Multiple Choice Rangers caught and banded 50 golden eagles at a wildlife reserve. Later, the researchers caught 200 golden eagles, 17 of which had bands. Estimate the total golden eagle population at the reserve.

Ⓐ 500 Ⓑ 526 Ⓒ 350

Ⓓ 588 Ⓔ 376

Quantitative Comparison In Exercises 9–11, solve the proportion. Then choose the statement below that is true about the given number.

Ⓐ The value in column A is greater.

Ⓑ The value in column B is greater.

Ⓒ The two values are equal

Ⓓ The relationship cannot be determined from the given information.

	Column A	Column B
9.	$\frac{3}{x} = \frac{10}{12}$	$\frac{2x+1}{14} = \frac{7}{9}$
10.	$\frac{x+7}{15} = \frac{x}{6}$	$\frac{x+2}{3} = \frac{1}{x}$
11.	$\frac{x+2}{10} = \frac{x}{x+3}$	$\frac{x-3}{x} = \frac{6}{2x+5}$

Standardized Test Practice

For use with pages 649–655

TEST TAKING STRATEGY Avoid spending too much time on one question. Skip questions that are too difficult for you and spend no more than a few minutes on each question.

1. *Multiple Choice*   What is 12% of 136?

Ⓐ 11.33   Ⓑ 16.32   Ⓒ 15.21

Ⓓ 14.7   Ⓔ 27.1

2. *Multiple Choice*   What is 420% of 48?

Ⓐ 87.5   Ⓑ 20.16   Ⓒ 100

Ⓓ 201.6   Ⓔ 875

3. *Multiple Choice*   13 is what percent of 57?

Ⓐ about 22.8%   Ⓑ about 7.41%

Ⓒ about 4.38%   Ⓓ about 438.5%

Ⓔ about 741%

4. *Multiple Choice*   75 is what percent of 20?

Ⓐ about 3.75%   Ⓑ about 26.7%

Ⓒ about 15%   Ⓓ about 267%

Ⓔ about 375%

5. *Multiple Choice*   31.5 is 15% of what number?

Ⓐ 200   Ⓑ 210   Ⓒ 230

Ⓓ 473   Ⓔ 476

6. *Multiple Choice*   90 is 120% of what number?

Ⓐ 108   Ⓑ 133   Ⓒ 75

Ⓓ 62   Ⓔ 85

7. *Multiple Choice*   A store has a 50% off sale in September on select items. In October, it marks the sale prices back up by 50%. What would a sweater that originally sold for $100 now be selling for?

Ⓐ $100   Ⓑ $50   Ⓒ $150

Ⓓ $25   Ⓔ $75

Multiple Choice   In Exercises 8 and 9, refer to the graph. In a survey, people were asked if they think the income tax system is equitable.

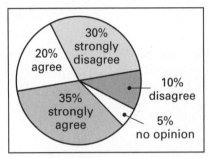

8. In the survey, 91 people strongly agreed that the income tax system is equitable. How many people were surveyed?

Ⓐ 250   Ⓑ 3185   Ⓒ 319

Ⓓ 455   Ⓔ 260

9. If you asked 550 people, how many would you expect to answer "disagree" or "strongly disagree"?

Ⓐ 165   Ⓑ 55   Ⓒ 220

Ⓓ 250   Ⓔ 303

Quantitative Comparison   In Exercises 10–13, solve the proportion. Then choose the statement below that is true about the given number.

Ⓐ The value in column A is greater.

Ⓑ The value in column B is greater.

Ⓒ The two values are equal

Ⓓ The relationship cannot be determined from the given information.

	Column A	Column B
10.	57% of 200 = x	110% of 60 = x
11.	38% of x = 540	19% of x = 270
12.	$\dfrac{17}{100} = \dfrac{x}{28}$	23% of x = 1.2
13.	x% of 72 = 18	20% of 125 = x

NAME _____ DATE _____

Standardized Test Practice

For use with pages 656–662

TEST TAKING STRATEGY   **Think positively during a test. This will help keep up your confidence and enable you to focus on each question.**

1. Multiple Choice   Which of the followed equations models inverse variation?

Ⓐ $y = \frac{1}{8}x$ Ⓑ $y = 8x$

Ⓒ $y = x + 8$ Ⓓ $y = 8 - x$

Ⓔ $xy = 8$

2. Multiple Choice   Which of the following equations models direct variation when $x = 7$ and $y = 21$?

Ⓐ $y = 3x$ Ⓑ $y = -3x$

Ⓒ $y = \frac{147}{x}$ Ⓓ $y = \frac{1}{3}x$

Ⓔ $y = \frac{x}{147}$

3. Multiple Choice   The variables x and y vary inversely. When x is 15, y is 40. If x is 5, then y is ___?___ .

Ⓐ 13.3   Ⓑ 115   Ⓒ 20

Ⓓ 120   Ⓔ 3000

4. Multiple Choice   The area of a rectangle is 15 square inches. Which equation shows the correct relationship of length to width?

Ⓐ $y = \frac{x}{15}$ Ⓑ $y = 15x$

Ⓒ $y = 15 - x$ Ⓓ $y = \frac{15}{x}$

Ⓔ $y = 15 + x$

5. Multiple Choice   Which of the following would *not* be an example of inverse variation?

Ⓐ The hours h you must work to earn $500 and your hourly pay p.

Ⓑ The base b and height h of a triangle if the area is 15 square inches.

Ⓒ The time t it takes to drive 30 miles and your rate r.

Ⓓ The rectangle's length l and the area h if the width is 5 inches.

Ⓔ None of these

6. Multiple Choice   Choose the type of variation and the equation that represents the graph.

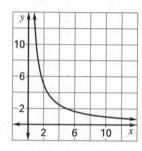

Ⓐ Inverse; $y = 10x$

Ⓑ Inverse; $y = \frac{10}{x}$

Ⓒ Direct; $y = 10x$

Ⓓ Direct; $y = \frac{10}{x}$

Ⓔ None of these

7. Multiple Choice   The variables x and y vary inversely. When x is -3, y is 18. If y is 6, then x is ___?___ .

Ⓐ -1 Ⓑ 1   Ⓒ -9

Ⓓ 9   Ⓔ 3

8. Multi-Step Problem   You would like to buy a new mountain bike that costs $1200.

a.   Make a table of possible values of the hours worked h and pay rate p to earn $1200. Sketch a graph.

b.   Does the model represent direct or inverse variation? Explain.

c.   *Critical Thinking*   Your friend is looking at a bike that costs $800. How does the model for this situation relate to the model for your situation?

NAME _____ DATE _____

Standardized Test Practice

For use with pages 664–669

TEST TAKING STRATEGY **Before you give up on a question, try to eliminate some of your choices so you can make an educated guess.**

1. *Multiple Choice* What is the simplified form of the expression $\dfrac{3x^2 + 6x}{x^2 + 9x + 14}$?

(A) $\dfrac{3}{x + 7}$ (B) $\dfrac{x + 2}{x + 7}$ (C) $\dfrac{3}{x + 2}$

(D) $\dfrac{3x}{x + 7}$ (E) $\dfrac{x + 3}{(x + 7)(x + 2)}$

2. *Multiple Choice* What is the simplified form of the expression $\dfrac{6x^2 + 3x - 9}{6x^2 + 45x + 54}$?

(A) $\dfrac{x - 3}{15x + 18}$ (B) $\dfrac{x - 1}{x + 6}$ (C) $\dfrac{3(x - 1)}{x + 6}$

(D) $\dfrac{3(2x + 3)}{x + 6}$ (E) $\dfrac{x - 1}{(x + 6)(2x + 3)}$

3. *Multiple Choice* What is the simplified form of the expression $\dfrac{2x^3 + 8x}{4x^3 - 16x}$?

(A) $\dfrac{x}{x + 2}$ (B) $\dfrac{1}{2}$ (C) $\dfrac{1}{2x^3 - 2}$

(D) $\dfrac{x}{2}$ (E) $\dfrac{x^2 + 4}{2(x + 2)(x - 2)}$

4. *Multiple Choice* Which values of the variable create an undefined expression for $\dfrac{12}{x^2 - x - 6}$?

(A) 2   (B) 2, −3   (C) 3

(D) −2, 3   (E) −2

5. *Multiple Choice* Which values of the variable create an undefined expression for $\dfrac{2x + 1}{2x^3 - 9x^2 - 5x}$?

(A) $5, -\tfrac{1}{2}$ (B) $-5, \tfrac{1}{2}$ (C) $0, -5$

(D) $0, 5, -\tfrac{1}{2}$ (E) $0, -5, \tfrac{1}{2}$

6. *Multiple Choice* What is the simplified form of the expression $\dfrac{-x^2 - 8x - 12}{x^2 + 2x - 24}$?

(A) $\dfrac{-4x - 1}{2}$ (B) $\dfrac{x + 2}{x - 4}$ (C) $\dfrac{-x - 2}{x - 4}$

(D) $\dfrac{-1}{x - 2}$ (E) $\dfrac{x - 6}{(x + 6)(x - 4)}$

7. *Multiple Choice* Which ratio represents the ratio of the smaller rectangle to the larger rectangle?

(A) $\dfrac{x}{x + 1}$ (B) $\dfrac{1}{3(x + 1)}$

(C) $\dfrac{x^2}{x^2 + 4x + 3}$ (D) $\dfrac{x + 3}{3x(x + 1)}$

(E) $\dfrac{x}{(x + 3)(x + 1)}$

8. *Multiple Choice* A coin is thrown onto the larger rectangle shown in Exercise 7. If it is equally likely to land anywhere in the rectangle, what is the probability that it will land on the small rectangle when $x = 5$?

(A) $\tfrac{1}{18}$ (B) $\tfrac{1}{5}$ (C) $\tfrac{13}{63}$

(D) $\tfrac{5}{48}$ (E) $\tfrac{13}{48}$

Quantitative Comparison In Exercises 9–11, simplify the expression and evaluate it for $x = 3$ and $y = -2$. Then choose the statement below that is true about the given values.

(A) The value in column A is greater.

(B) The value in column B is greater.

(C) The two values are equal

(D) The relationship cannot be determined from the given information.

	Column A	Column B
9.	$\dfrac{5x}{x^2 + 6x}$	$\dfrac{7y^2}{14y - 7y^2}$
10.	$\dfrac{x^2 - 4}{x^2 + 4x + 4}$	$\dfrac{y^2 + 9}{y^2 + 6y + 9}$
11.	$\dfrac{2x - 6}{x^2 - x - 6}$	$\dfrac{y^3 + 2y^2}{y^4 - 3y^3 - 2y^2}$

NAME _____   DATE _____

Standardized Test Practice

For use with pages 670–675

TEST TAKING STRATEGY   **Go back and check as much of your work as you can.**

1. **Multiple Choice**   What is the simplified form of the expression $\dfrac{5x^3}{7x^2} \cdot \dfrac{21x^2}{20x}$?

   (A) $\dfrac{3x^3}{4}$ **(B)** $\dfrac{x^2}{2}$ **(C)** $\dfrac{x^3}{4}$

   (D) $\dfrac{3x^3}{4x}$ **(E)** $\dfrac{3x^2}{4}$

2. **Multiple Choice**   Which product equals the quotient $(5x - 1) \div \dfrac{x^2 + 4}{x - 1}$?

   (A) $\dfrac{1}{(5x - 1)} \cdot \dfrac{x^2 + 4}{x + 1}$

   (B) $\dfrac{(5x - 1)}{1} \cdot \dfrac{x - 1}{x^2 + 4}$

   (C) $\dfrac{(5x - 1)}{1} \cdot \dfrac{x^2 + 4}{x - 1}$

   (D) $\dfrac{1}{(5x - 1)} \cdot \dfrac{x + 1}{x^2 + 4}$

   (E) $\dfrac{5x^2 - 1}{x - 1} \cdot \dfrac{x^2 + 4}{x - 1}$

3. **Multiple Choice**   What is the simplified form of the expression

   $\dfrac{x + 3}{2x^2 + x - 10} \cdot \dfrac{2x + 5}{2x + 6}$?

   (A) $\dfrac{1}{2(x - 2)}$ **(B)** $\dfrac{1}{2x + 5}$

   (C) $\dfrac{x + 3}{2(x - 2)}$ **(D)** $\dfrac{2x + 5}{2(x + 3)}$

   (E) $\dfrac{x + 3}{(x - 2)(2x + 6)}$

4. **Multiple Choice**   What is the simplified form of the expression $\dfrac{x^2 - 1}{x^2 + 6x} \div \dfrac{x - 1}{7x^2}$?

   (A) $\dfrac{x - 1}{7x(x + 6)}$ **(B)** $\dfrac{7x(x + 1)}{x + 6}$

   (C) $\dfrac{x + 6}{7(x + 1)}$ **(D)** $\dfrac{x + 1}{7(x + 6)}$

   (E) $\dfrac{6x^2 - 1}{x^2 + 5x - 1}$

5. **Multiple Choice**   What is the simplified form of the expression

   $\dfrac{5x^2 + 23x - 42}{x + 7} \div (5x - 7)$?

   (A) $\dfrac{x + 7}{(x + 6)(5x - 7)}$ **(B)** $\dfrac{x + 7}{x + 6}$

   (C) $\dfrac{5x - 7}{x + 6}$ **(D)** $\dfrac{x + 6}{x + 7}$

   (E) $\dfrac{(x + 6)(5x - 7)^2}{x + 7}$

6. **Multiple Choice**   What is the simplified form of the expression

   $\left(\dfrac{x^2}{5x + 1} \cdot \dfrac{(x + 1)}{3x^3}\right) \div \dfrac{x}{10x + 2}$?

   (A) $\dfrac{x(x + 1)}{5x + 1}$ **(B)** $\dfrac{2x}{3(5x + 1)}$

   (C) $\dfrac{2(x + 1)}{3x^2}$ **(D)** $\dfrac{2(5x + 1)}{3x^2}$

   (E) $\dfrac{x + 1}{3x^2}$

7. **Multi-Step Problem**   A local trucking company kept data on its company from 1995–1998 and developed the following models, where t is the number of years since 1995.

   Number of miles driven:
   $$M = \dfrac{2{,}000{,}000 + 23{,}000t}{1 - 0.03t}$$

   Number of employees: $E = \dfrac{20 + 5t}{1 + 0.02t}$

   a. Find a model for the average number of miles driven per employee.

   b. Use the model to predict the average number of miles driven in 2002. Round your answer to the nearest hundred miles.

   c. Use the model to predict the average number miles driven in 2004. Round your answer to the nearest hundred miles.

   d. *Writing*   Compare the answers to parts (b) and (c). What is the trend? Can this continue indefinitely?

NAME _____ DATE _____

Standardized Test Practice

For use with pages 676–682

TEST TAKING STRATEGY   **Read all of the answer choices before deciding which is the correct one.**

1. *Multiple Choice*   Simplify the expression

$$\frac{7x + 2}{4x + 3} + \frac{6x}{4x + 3}.$$

(A) $x + 2$ **(B)** $\dfrac{x + 2}{4x + 3}$ **(C)** $\dfrac{13x + 2}{8x + 6}$

(D) $\dfrac{13x + 2}{4x + 3}$ **(E)** $\dfrac{42x^2 + 12x}{4x + 3}$

2. *Multiple Choice*   Find the LCD of

$$\frac{x}{x^2 - x - 6} \text{ and } \frac{3x + 1}{x^2 - 6x + 9}.$$

(A) $(x + 2)(x - 3)$ **(B)** $(x - 3)(x - 3)$

(C) $x(3x + 1)$ **(D)** $(x - 2)(3x + 1)$

(E) $(x + 2)(x - 3)^2$

3. *Multiple Choice*   Simplify the expression

$$\frac{2}{x + 2} + \frac{5}{x + 3} + \frac{x + 1}{x^2 + 5x + 6}.$$

(A) $\dfrac{7x + 7}{(x + 2)(x + 3)}$ **(B)** $\dfrac{x + 8}{(x + 2)(x + 3)}$

(C) $\dfrac{x + 8}{x + 3}$ **(D)** $\dfrac{8x + 17}{(x + 2)(x + 3)}$

(E) $\dfrac{8x + 17}{(x + 2)^2(x + 3)^2}$

4. *Multiple Choice*   Simplify the expression

$$\frac{3x}{x - 1} - \frac{x - 1}{x + 5}.$$

(A) $\dfrac{3x^2 + 15x}{x + 5}$ **(B)** $\dfrac{2x^2 + 17x - 1}{(x + 5)(x - 1)}$

(C) $\dfrac{3x^2 + 15x - 1}{x + 5}$ **(D)** $\dfrac{2x^2 - 2x + 16}{(x + 5)(x - 1)}$

(E) $\dfrac{2x + 1}{(x + 5)(x - 1)}$

5. *Multiple Choice*   Which expression represents perimeter of the rectangle?

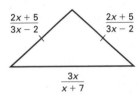

(A) $\dfrac{13x^2 + 32x + 70}{(3x - 2)(x + 7)}$

(B) $\dfrac{4x^2 + 42x + 68}{(3x - 2)(x + 7)}$

(C) $\dfrac{13x^2 + 14x + 25}{(3x - 2)(x + 7)}$

(D) $\dfrac{13x^2 + 44x + 70}{(x + 7)(3x - 2)}$

(E) $\dfrac{4x^2 + 42x + 68}{(3x - 2)^2(x + 7)}$

6. *Multi-Step Problem*   A plane is traveling 1200 miles between two cities. Its speed with no wind is represented by x miles per hour. On a trip, traveling with the wind, the wind speed is 25 miles per hour. On the return trip, traveling against the wind, the wind speed increases to 30 miles per hour.

   a. Write an expression for the total time for the round trip.

   b. Simplify the expression from part (a).

   c. How long will the trip take, both ways, if the plane travels 200 miles per hour?

   d. *Critical Thinking*   A smaller plane can only travel 100 miles per hour. Compare the travel time of the planes. Did you get the result you expected? Explain.

NAME _____ DATE _____

Standardized Test Practice

For use with pages 684–689

TEST TAKING STRATEGY Be aware of how much time you have left, but keep focused on your work.

1. **Multiple Choice** What is $12x^2 - 4x + 7$ divided by $-2x$?

   Ⓐ $-6x + 2 + \dfrac{7}{2x}$ Ⓑ $6x - 2 + \dfrac{7}{2x}$

   Ⓒ $-6x + 2 + \dfrac{7}{-2x}$ Ⓓ $6x + 2 + \dfrac{7}{2x}$

   Ⓔ $-6x - 2 + \dfrac{7}{-2x}$

2. **Multiple Choice** What is $21x^2 + 17x - 5$ divided by $7x + 1$?

   Ⓐ $3x + 3 + \dfrac{2x - 5}{7x + 1}$

   Ⓑ $3x + 2 + \dfrac{-7}{7x + 1}$

   Ⓒ $3x + 3 + \dfrac{-2x + 5}{7x + 1}$

   Ⓓ $3x + 2 + \dfrac{-3}{7x + 1}$

   Ⓔ $14x + 10 + \dfrac{-5}{7x + 1}$

3. **Multiple Choice** In the equation $\dfrac{5x^2 + 2x + 3}{x + 2} = 5x - 8 + \dfrac{19}{x + 2}$, the $x + 2$ represents the ___?___ .

   Ⓐ dividend   Ⓑ quotient

   Ⓒ remainder   Ⓓ divisor

   Ⓔ product

4. **Multiple Choice** What is $15x^2 - 11x - 6$ divided by $5x - 2$?

   Ⓐ $3x - 1 + \dfrac{-4}{5x - 2}$

   Ⓑ $3x - 3 + \dfrac{-2x - 6}{5x - 2}$

   Ⓒ $3x - 3 + \dfrac{2x - 6}{5x - 2}$

   Ⓓ $3x - 1 + \dfrac{-8}{5x - 2}$

   Ⓔ $3x - 5 + \dfrac{7}{5x - 2}$

5. **Multiple Choice** The area of a rectangle is $15x^2 - x - 28$. One side is $3x + 4$. Choose the expression for the missing data.

   Ⓐ $5x + 6 + \dfrac{-4}{3x + 4}$

   Ⓑ $5x + 7$

   Ⓒ $5x - 7 + \dfrac{1}{3x + 4}$

   Ⓓ $5x + 6 + \dfrac{-52}{3x + 4}$

   Ⓔ $5x - 7$

6. **Multiple Choice** Jane is twice as old as Mitch. Write and simplify using long division a ratio comparing Jane's age in 6 years to Mitch's age in 6 years.

   Ⓐ $2 + \dfrac{18}{x + 6}$ Ⓑ $2 - \dfrac{6}{x + 6}$

   Ⓒ $2 + \dfrac{6}{x + 6}$ Ⓓ $2 - \dfrac{18}{x + 6}$

   Ⓔ $\dfrac{2}{x + 6}$

Quantitative Comparison In Exercises 7–9, use long division to divide the polynomials. Then evaluate the expression when $x = 2$ and $y = 3$. Choose the statement below that is true about the given values.

   Ⓐ The value in column A is greater.

   Ⓑ The value in column B is greater.

   Ⓒ The two values are equal.

   Ⓓ The relationship cannot be determined from the given information.

	Column A	Column B
7.	$(12x^3 + 4x^2) \div 4x$	$(18y^3 - 9y^2 - 3) \div 3y$
8.	$(10x^2 - 5x - 7) \div (2x - 4)$	$(5y^2 + 8y + 1) \div (y + 3)$
9.	$(6x^2 + 7x - 3) \div (2x + 3)$	$(15y^2 + 16y + 4) \div (5y + 2)$

TEST TAKING STRATEGY   **As soon as the testing begins, start working. Keep moving and stay focused on the test.**

1. *Multiple Choice* What is the solution of $\dfrac{x+1}{5} = \dfrac{2x}{15}$?

- **A** $-\dfrac{3}{5}$
- **B** -3
- **C** 3
- **D** $-\dfrac{1}{5}$
- **E** $\dfrac{3}{5}$

2. *Multiple Choice* What is a solution of $\dfrac{12}{x+2} + 2 = \dfrac{3x}{x^2 - 3x - 10}$?

- **A** about 7.1
- **B** about 5.6
- **C** about -5.6
- **D** about -4.4
- **E** about 2.9

3. *Multiple Choice* You have a batting average of 0.200 after 90 times at bat. You would like to raise your average to 0.250. Which equation would you use to calculate the number of consecutive hits you need to achieve your goal?

- **A** $0.25 = \dfrac{18+x}{90+x}$
- **B** $0.25 = \dfrac{18+x}{90}$
- **C** $0.20 = \dfrac{18+x}{90+x}$
- **D** $0.20 = \dfrac{x-18}{90}$
- **E** $0.05 = \dfrac{x}{18}$

4. *Multiple Choice* Two hoses are filling a swimming pool. The first hose can fill the pool in 30 minutes, the second in 45 minutes. If both hoses are used, how long will it take to fill the pool?

- **A** 15 minutes
- **B** 16 minutes
- **C** 18 minutes
- **D** 20 minutes
- **E** 22 minutes

5. *Multiple Choice* Which value is not in the domain of $y = \dfrac{1}{x-5} + 6$?

- **A** -5
- **B** -6
- **C** 0
- **D** 6
- **E** 5

6. *Multiple Choice* Which function represents the graph?

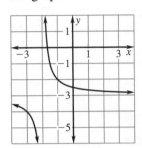

- **A** $y = \dfrac{1}{x+2} + 3$
- **B** $y = \dfrac{1}{x+2} - 3$
- **C** $y = \dfrac{-1}{x+2} + 3$
- **D** $y = \dfrac{-1}{x+2} - 3$
- **E** $y = \dfrac{1}{x-3} - 1$

11. *Multiple Choice* Choose the equation of the vertical asymptote of the hyperbola $y = \dfrac{-4}{x-5} + 3$.

- **A** $y = 5$
- **B** $y = -5$
- **C** $x = 3$
- **D** $x = -3$
- **E** $x = 5$

Quantitative Comparison In Exercises 8–10, find the center of the hyperbola, then choose the statement below that is true about the given statement.

- **A** The value of h is greater than k.
- **B** The value of k is greater than h.
- **C** The values of h and k are equal.
- **D** The relationship cannot be determined from the information given.

8. $y = \dfrac{-7}{x+3} - 4$

9. $y = \dfrac{5x+2}{x-1}$

10. $y = \dfrac{8x-35}{x-5}$

Chapter 11

TEST TAKING STRATEGY **If you can, check your answer using a different method than you used originally, to avoid making the same mistake twice.**

1. *Multiple Choice*   What is the domain of the function $y = 5 + \sqrt{x + 6}$?

   Ⓐ $x \geq -6$ Ⓑ $x \leq -6$ Ⓒ $x \geq 6$

   Ⓓ $x \leq 6$ Ⓔ $x \geq 0$

2. *Multiple Choice*   Which quadrants of the coordinate plane will contain the graph of $y = \sqrt{2x - 4}$?

   Ⓐ Quadrant I   Ⓑ Quadrant III

   Ⓒ Quadrant IV   Ⓓ Quadrant I and IV

   Ⓔ Quadrant III and IV

3. *Multiple Choice*   What are the domain and the range of the function $y = \sqrt{x - 2} - 5$?

   Ⓐ Domain: $x \geq 0$ Range: $y \geq 0$

   Ⓑ Domain: $x \geq 2$ Range: $y \geq -5$

   Ⓒ Domain: $x \geq 5$ Range: $y \geq -2$

   Ⓓ Domain: $x \geq -2$ Range: $y \geq 0$

   Ⓔ Domain: $x \geq 2$ Range: $y \geq 5$

4. *Multiple Choice*   What is the equation of the graph of the function shown?

   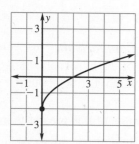

   Ⓐ $y = \sqrt{2x} - 2$ Ⓑ $y = \sqrt{x} - 2$

   Ⓒ $y = \sqrt{x} + 2$ Ⓓ $y = \sqrt{2x - 2}$

   Ⓔ $y = \sqrt{x - 2}$

5. *Multiple Choice*   What is the walking speed of an animal with a leg length of 3 feet. The walking speed model is $s = \sqrt{gL}$, where $g = 32$ ft/sec²

   Ⓐ about 4.8 ft/sec   Ⓑ about 3.3 ft/sec

   Ⓒ about 9.8 ft/sec   Ⓓ about 5.9 ft/sec

   Ⓔ about 6.6 ft/sec

6. *Multiple Choice*   What is the radius of a cylindrical container with a volume of 196.25 cubic inches. The height is 10 inches. (*Hint:* $V = \pi r^2 h$.)

   Ⓐ about 2.2 inches   Ⓑ about 10.3 inches

   Ⓒ about 2.5 inches   Ⓓ about 13.5 inches

   Ⓔ about 12.8 inches

Quantitive Comparison   In Exercises 7–9, evaluate each the square root function for the given values of x, then choose the statement that is true about the given numbers.

Ⓐ The value in column A is greater.

Ⓑ The value in column B is greater.

Ⓒ The two values are equal.

Ⓓ The relationship cannot be determined from the given information.

	$x =$	Column A	Column B
7.	When $x = 3$	$y = \dfrac{1}{3}\sqrt{3x} - 1$	$y = \sqrt{6 - 2x}$
8.	When $x = -2$	$y = \sqrt{6 - 5x}$	$y = \sqrt{4x^2 + 9}$
9.	When $x = 5$	$y = \sqrt{\dfrac{x}{4} - 1}$	$y = \sqrt{\dfrac{x}{3} + 1}$

NAME _____ DATE _____

Standardized Test Practice

For use with pages 716–721

TEST TAKING STRATEGY Work as fast as you can through the easier problems, but not so fast that you are careless.

1. *Multiple Choice* Simplify the expression $\sqrt{500} - \sqrt{180} + \sqrt{80}$.

Ⓐ $12\sqrt{2}$ Ⓑ $5\sqrt{8}$ Ⓒ $20\sqrt{5}$

Ⓓ $8\sqrt{5}$ Ⓔ 20

2. *Multiple Choice* The conjugate of $5 - \sqrt{3}$ is __?__.

Ⓐ $5 + \sqrt{3}$ Ⓑ $5 - \sqrt{3}$

Ⓒ $5 + \sqrt{-3}$ Ⓓ $-5 - \sqrt{3}$

Ⓔ $-5 + \sqrt{3}$

3. *Multiple Choice* Simplify the expression $\dfrac{18}{4 - \sqrt{7}}$.

Ⓐ $72 + 18\sqrt{7}$ Ⓑ $72 - 18\sqrt{7}$

Ⓒ $\dfrac{72 - 18\sqrt{7}}{9}$ Ⓓ $\dfrac{72 - 18\sqrt{7}}{9}$

Ⓔ $8 + 2\sqrt{7}$

4. *Multiple Choice* Solve the quadratic equation $x^2 - 5x - 2 = 0$.

Ⓐ $\dfrac{-5 \pm \sqrt{33}}{2}$ Ⓑ $\dfrac{5 \pm \sqrt{33}}{2}$

Ⓒ $\dfrac{5 \pm \sqrt{17}}{2}$ Ⓓ $\dfrac{-5 \pm \sqrt{17}}{2}$

Ⓔ $\dfrac{5 \pm \sqrt{31}}{2}$

5. *Multiple Choice* Find the area of the triangle below.

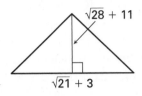

Ⓐ $14\sqrt{3} + 17\sqrt{7} + 33$

Ⓑ $7\sqrt{6} + 11\sqrt{21} + 6\sqrt{7} + 33$

Ⓒ $14\sqrt{3} + 11\sqrt{21} + 6\sqrt{7} + 33$

Ⓓ $2\sqrt{39} + 11\sqrt{21} + 7\sqrt{6} + 33$

Ⓔ $24\sqrt{4} + 3\sqrt{28} + 11\sqrt{21} + 33$

6 *Multiple Choice* The area of a rectangle is $5 + \sqrt{3}$. If the length of one side is $2 + \sqrt{3}$, what is the length of the other side?

Ⓐ $13 + 3\sqrt{3}$ Ⓑ $7 + 7\sqrt{3}$

Ⓒ $13 - 3\sqrt{3}$ Ⓓ $7 + 3\sqrt{3}$

Ⓔ $7 - 3\sqrt{3}$

Quantitative Comparison In Exercises 7–9, simplify the expression. Then choose the statement below that is true about the given number.

Ⓐ The value in column A is greater.

Ⓑ The value in column B is greater.

Ⓒ The two values are equal.

Ⓓ The relationship cannot be determined from the given information.

	Column A	*Column B*
7.	$\left(2 + \sqrt{6}\right)\left(2 - \sqrt{6}\right)$	$2\sqrt{3} \cdot \sqrt{12}$
8.	$\sqrt{75} - \sqrt{27}$	$\sqrt{72} - \sqrt{18}$
9.	$\dfrac{3}{\sqrt{5}} \cdot \dfrac{2\sqrt{15}}{3}$	$\dfrac{\sqrt{12}}{3} \cdot \dfrac{\sqrt{18}}{\sqrt{2}}$

Chapter 12

TEST TAKING STRATEGY   **Learn as much as you can about a test ahead of time, such as the types of questions and the topics that the test will cover.**

1. *Multiple Choice*   Which of the following is a solution of $\sqrt{x} - 5 = 10$?

   Ⓐ 15   Ⓑ 30   Ⓒ 5.5

   Ⓓ 225   Ⓔ 230

2. *Multiple Choice*   Which of the following is a solution of $\sqrt{2x + 3} - 3 = 8$?

   Ⓐ 118   Ⓑ 62   Ⓒ 59

   Ⓓ 32   Ⓔ 31

3. *Multiple Choice*   Which of the following is a solution of $3 - \sqrt{2x - 1} = 5$?

   Ⓐ $\frac{2}{5}$ Ⓑ $\frac{5}{2}$ Ⓒ $\frac{3}{5}$

   Ⓓ $\frac{5}{3}$ Ⓔ $\frac{3}{2}$

4. *Multiple Choice*   Which of the following is a solution of $x = \sqrt{4x + 32}$?

   Ⓐ 8   Ⓑ 4   Ⓒ −4

   Ⓓ 8, 4   Ⓔ 8, −4

5. *Multiple Choice*   Which of the following is a solution of $x = \sqrt{2x + 48}$?

   Ⓐ 6   Ⓑ 8   Ⓒ −24

   Ⓓ −6   Ⓔ −8

6. *Multiple Choice*   The geometric mean of x and 27 is 18. What is the value of x?

   Ⓐ 22   Ⓑ 13   Ⓒ 12

   Ⓓ 21   Ⓔ 16

7. *Multiple Choice*   The geometric mean of 12 and y is 42. What is the value of y?

   Ⓐ 42   Ⓑ 22.4   Ⓒ 3.4

   Ⓓ 3.5   Ⓔ 147

8. *Multiple Choice*   What is the value of x in the triangle below whose area is 26?

   Ⓐ 1

   Ⓑ 2

   Ⓒ 3

   Ⓓ 4

   Ⓔ 5

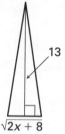

13

$\sqrt{2x + 8}$

9. *Multiple Choice*   Two numbers have a geometric mean of 6. One number is 9 less than other. What are the numbers?

   Ⓐ 3 and 12, −3 and −12

   Ⓑ 2 and 11, −2 and −11

   Ⓒ 4 and 13, −4 and −13

   Ⓓ 1 and 10, −1 and −10

   Ⓔ None of these

Quantitative Comparison   In Exercises 10–13, choose the statement below that is true about the given quantities.

   Ⓐ The quantity in column A is greater.

   Ⓑ The quantity in column B is greater.

   Ⓒ The two quantity are equal.

   Ⓓ The relationship cannot be determined from the given information.

	Column A	Column B
10.	The geometric mean of 4 and 9.	The geometric mean of 2 and 18.
11.	The solution of $\sqrt{x} + 5 = 3$.	The solution of $\sqrt{x + 5} = 3$.
12.	The geometric mean of 3 and 75.	The geometric mean of 12 and 27.
13.	The geometric mean of −4 and −16.	The solution of $\sqrt{2x + 7} = 4$.

NAME _____ DATE _____

Standardized Test Practice

For use with pages 730–736

TEST TAKING STRATEGY   Avoid spending too much time on one question. Skip questions that are too difficult for you, and spend no more than a few minutes on each question.

1. *Multiple Choice*   What term should be added to $x^2 + 18x$ so that the result is a perfect square trinomial?

   (A) 9   (B) 162   (C) 42

   (D) 81   (E) 36

2. *Multiple Choice*   What term should be added to $x^2 - \frac{6}{5}x$ so that the result is a perfect square trinomial?

   (A) $\frac{3}{5}$ (B) $\frac{144}{25}$ (C) $\frac{9}{25}$

   (D) $\frac{12}{5}$ (E) $-\frac{3}{5}$

3. *Multiple Choice*   Solve $4x^2 - x - 8 = 0$ by completing the square, then choose the solution.

   (A) $\frac{1}{8} \pm \frac{\sqrt{130}}{8}$ (B) $\frac{1}{8} \pm \frac{\sqrt{129}}{8}$

   (C) $\frac{1}{4} \pm \frac{\sqrt{129}}{8}$ (D) $-\frac{1}{8} \pm \frac{\sqrt{129}}{8}$

   (E) None of these

4. *Multiple Choice*   Solve $x^2 - 12x - 7 = 0$ by completing the square.

   (A) $7 \pm \sqrt{43}$ (B) $6 \pm \sqrt{43}$

   (C) $6 \pm \sqrt{42}$ (D) $7 \pm \sqrt{42}$

   (E) $12 \pm \sqrt{43}$

5. *Multiple Choice*   Solve $15x - 45x^2 = 0$ using the most appropriate method.

   (A) $0, -\frac{1}{3}$ (B) $0, 3$ (C) $0, \frac{1}{3}$

   (D) $1, \frac{1}{3}$ (E) $0, -3$

6. *Multiple Choice*   Solve the equation $3x^2 + 13x - 10 = 0$ using the most appropriate method.

   (A) $-\frac{2}{3}, 5$ (B) $-5, \frac{2}{3}$ (C) $-5, \frac{3}{2}$

   (D) $10, \frac{5}{3}$ (E) $-10, \frac{5}{3}$

7. *Multiple Choice*   The area of a triangle is 90 square inches. The base is 8 inches less than 3 times the height. What is the measurement of the height?

   (A) 6.5 inches   (B) 7.0 inches

   (C) 8.2 inches   (D) 9.2 inches

   (E) 4.3 inches

8. *Multiple Choice*   Solve $x^2 - 14x - 11 = 0$ using the most appropriate method.

   (A) $7 \pm 6\sqrt{10}$ (B) $-7 \pm 3\sqrt{10}$

   (C) $-7 \pm 2\sqrt{5}$ (D) $-7 \pm 3\sqrt{5}$

   (E) $7 \pm 2\sqrt{15}$

Quantitative Comparison   In Exercises 9–11, find the term that should be added to the expression to create a perfect square trinomial. Then choose the statement below that is true about the added term.

   (A)   The added term in column A is greater.

   (B)   The added term in column B is greater.

   (C)   The two added terms are equal.

   (D)   The relationship cannot be determined from the given information.

	Column A	Column B
9.	$x^2 + 13x$	$x^2 - 13x$
10.	$x^2 + \frac{3}{5}x$	$x^2 - \frac{4}{7}x$
11.	$x^2 + 3.2x$	$x^2 + \frac{17}{5}x$

TEST TAKING STRATEGY   Be aware of how much time you have left, but keep focused on your work.

1. *Multiple Choice*   What is the length of the missing side of the triangle?

  Ⓐ $6\sqrt{3}$ Ⓑ $18\sqrt{2}$

  Ⓒ $3\sqrt{6}$ Ⓓ $4\sqrt{3}$

  Ⓔ $3\sqrt{4}$

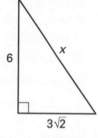

2. *Multiple Choice*   What is the length of the missing side of the triangle?

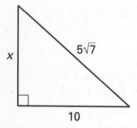

  Ⓐ $5\sqrt{3}$ Ⓑ $7\sqrt{2}$

  Ⓒ $11\sqrt{5}$ Ⓓ $3\sqrt{5}$

  Ⓔ $5\sqrt{11}$

3. *Multiple Choice*   A right triangle has one leg that is 5 inches longer than the other leg. The hypotenuse is 25. Find the length of the shorter leg.

  Ⓐ 10   Ⓑ 15   Ⓒ 17

  Ⓓ 20   Ⓔ 13

4. *Multiple Choice*   A right triangle has one leg that is twice the other leg. If the hypotenuse is $8\sqrt{5}$, find the length of the longer leg.

  Ⓐ 8   Ⓑ 16   Ⓒ 4

  Ⓓ 32   Ⓔ 12

5. *Multiple Choice*   Which lengths below would form a right triangle?

  Ⓐ 1, 3, 5   Ⓑ 1.2, 2.4, 6.1

  Ⓒ 6, 9, 10   Ⓓ $2\sqrt{5}$, 4, 6

  Ⓔ None of these

6. *Multiple Choice*   A rectangular pool is 13 feet by 25 feet. What is the diagonal length from corner to corner?

  Ⓐ about 29.5   Ⓑ about 27 ft

  Ⓒ about 27.5   Ⓓ about 28.5

  Ⓔ about 28 ft

7. *Multiple Choice*   The area of a square is 64 square feet. What is the length of its diagonal?

  Ⓐ 8 feet   Ⓑ $8\sqrt{3}$ feet

  Ⓒ $9\sqrt{2}$ feet   Ⓓ $8\sqrt{2}$ feet

  Ⓔ $5\sqrt{6}$ feet

8. *Multiple Choice*   The diagonal of a square is $5\sqrt{2}$ feet. What is its area?

  Ⓐ 5 ft^2 Ⓑ 10 ft^2

  Ⓒ 15 ft^2 Ⓓ 25 ft^2

  Ⓔ 50 ft^2

9. *Mulit-Step Problem*   Two planes leave from Philadelphia airport at the same time. One flies due south. The other flies southwest at a rate that is 30 miles per hour less than twice the rate of the plane headed south. After one hour, measuring east to west, the planes are 200 miles apart.

  a. Let x represent the plane headed South. Write an expression for each plane's distance after one hour.

  b. Draw a diagram of the situation.

  c. Use the Pythagorean theorem to find the speed of each plane.

  d. *Writing*   Which method did you use to solve the quadratic equation? Why?

Chapter 12

NAME _____ DATE _____

Standardized Test Practice

For use with pages 745–750

TEST TAKING STRATEGY   **Think positively during a test. This will help keep your confidence and enable you to focus on each question.**

1. *Multiple Choice*   What is the distance between $(-3, -2)$ and $(-4, 3)$?

   Ⓐ $\sqrt{74}$ Ⓑ $\sqrt{2}$ Ⓒ $\sqrt{34}$

   Ⓓ $\sqrt{26}$ Ⓔ $\sqrt{8}$

2. *Multiple Choice*   What is the distance between $(3, 5)$ and $(8, -7)$?

   Ⓐ $\sqrt{29}$ Ⓑ 13 Ⓒ $\sqrt{109}$

   Ⓓ 5 Ⓔ $\sqrt{265}$

3. *Multiple Choice*   What is the distance between between points A and B?

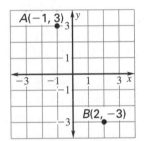

   Ⓐ $3\sqrt{5}$ Ⓑ 3 Ⓒ $\sqrt{10}$

   Ⓓ $3\sqrt{10}$ Ⓔ $\sqrt{5}$

4. *Multiple Choice*   Use the graph in Exercise 3. Find the midpoint between points *A* and *B*.

   Ⓐ $\left(0, \frac{1}{2}\right)$ Ⓑ $\left(-\frac{3}{4}, -3\right)$

   Ⓒ $\left(-\frac{1}{2}, 0\right)$ Ⓓ $\left(\frac{1}{2}, 1\right)$

   Ⓔ $\left(\frac{1}{2}, 0\right)$

5. *Multiple Choice*   The vertices of a right triangle are $(5, -2)$, $(0, 6)$, and $(0, -2)$. What is the length of the hypotenuse?

   Ⓐ 8 Ⓑ 4 Ⓒ $\sqrt{89}$

   Ⓓ $\sqrt{41}$ Ⓔ 5

6. *Multiple Choice*   The vertices of a right triangle are $(-1, -5)$, $(3, 2)$, and $(3, -5)$. What is the length of the hypotenuse?

   Ⓐ 4 Ⓑ 7 Ⓒ $\sqrt{65}$

   Ⓓ $\sqrt{13}$ Ⓔ $3\sqrt{5}$

7. *Multiple Choice*   What is the midpoint between $(10, 3)$ and $(-6, 2)$?

   Ⓐ $\left(8, -\frac{1}{2}\right)$ Ⓑ $\left(2, \frac{5}{2}\right)$ Ⓒ $\left(8, -\frac{1}{2}\right)$

   Ⓓ $\left(-2, -\frac{5}{2}\right)$ Ⓔ $\left(-2, -\frac{1}{2}\right)$

8. *Multiple Choice*   Which set of points below form a right triangle?

   Ⓐ $(3, 2), (5, 7), (-1, 7)$

   Ⓑ $(-2, 1), (4, 8), (6, 1)$

   Ⓒ $(-1, 2), (1, 2), (6, -1)$

   Ⓓ $(-5, -3), (4, -3), (-5, 4)$

   Ⓔ   None of these

9. *Multiple Choice*   Two fish are stocked into a lake from a boat. The first swims 1 mile north and 2 miles east. The second swims 3 miles west then 2 miles north. How many miles apart are the fish?

   Ⓐ 5.8 Ⓑ 4 Ⓒ 4.5

   Ⓓ 5.1 Ⓔ 5.6

10. *Multiple Choice*   The distance between two points is 7, and the points are $(5, 2)$ and $(x, 4)$. Find x.

   Ⓐ 12.3 Ⓑ -2.3 Ⓒ -3.2

   Ⓓ   A and B   Ⓔ   B and C

Quantitative Comparison   In Exercises 11–13, find the distance between the two points. Then choose the statement below that is true about the given number.

   Ⓐ   The value in column A is greater.

   Ⓑ   The value in column B is greater.

   Ⓒ   The two values added term are equal.

   Ⓓ   The relationship cannot be determined from the given information.

	Column A	Column B
11.	$(5, 8)$ and $(2, 1)$	$(-4, 2)$ and $(3, 2)$
12.	$(-2, -2)$ and $(-5, -13)$	$(4, -2)$ and $(7, 9)$
13.	$\left(\frac{1}{2}, 6\right)$ and $\left(2, \frac{2}{3}\right)$	$\left(\frac{5}{4}, -2\right)$ and $\left(3, -\frac{1}{2}\right)$

Chapter 12

NAME _____ DATE _____

Standardized Test Practice

For use with pages 752–757

TEST TAKING STRATEGY **Read all of the answer choices before deciding which is the correct one.**

1. *Multiple Choice* What is the sine of $\angle A$ in the triangle?

A $\frac{12}{13}$

B $\frac{5}{12}$

C $\frac{5}{13}$

D $\frac{13}{12}$

E $\frac{12}{5}$

2. *Multiple Choice* Use the figure in Exercise 1. What is the cosine of $\angle B$ in the triangle?

A $\frac{12}{13}$ **B** $\frac{5}{13}$ **C** $\frac{12}{5}$

D $\frac{5}{12}$ **E** $\frac{13}{12}$

3. *Multiple Choice* Use the figure in Exercise 1. What is the tangent of $\angle A$ in the triangle?

A $\frac{5}{13}$ **B** $\frac{12}{5}$ **C** $\frac{12}{13}$

D $\frac{13}{12}$ **E** $\frac{5}{12}$

4. *Multiple Choice* What is the length of side a in the triangle?

A 12.5

B 14.4

C 20.0

D 21.65

E 34.64

5. *Multiple Choice* Use the figure in Exercise 4. What is the length of side c in the triangle?

A 12.5   **B** 14.4   **C** 20.0

D 34.64   **E** 21.65

6. *Multiple Choice* Use the figure in Exercise 7. What is the length of the hypotenuse in the triangle?

A 9.58

B 13.27

C 15.92

D 7.22

E 24

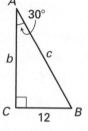

7. *Multiple Choice* What is the length of side b in the triangle?

A 24.26   **B** 27   **C** 24

D 9   **E** 20.78

8. *Multi-Step Problem* Global positioning systems are devices that can pin-point a location on Earth by use of satelite systems and triangulation. By bouncing your signal off of two satellites with known coordinates, you can always find your position.

A satellite is 1500 miles directly above you and a second is 700 miles from the first satellite. If you are facing due east, the second satellite is at an angle of 60° from the horizon. Assume the two satellites and your position form a right triangle.

a. Draw a diagram of the situation.

b. Find the distance between you and the second satellite.

c. *Critical Thinking* Can you always use the Pythagorean theorem to solve a global positioning problem? Explain.

Chapter 12

NAME _____ DATE _____

Standardized Test Practice

For use with pages 758–764

TEST TAKING STRATEGY   **As soon as the testing begins, start working. Keep moving and stay focused on the test.**

1. *Multiple Choice*   The basic axiom of algebra represented by $x\left(\dfrac{1}{x}\right) = 1$ where x is any real number not equal to zero is ___?___.

   Ⓐ   commutative property of multiplication
   Ⓑ   associative property of multiplication
   Ⓒ   identity property of multiplication
   Ⓓ   inverse property of multiplication
   Ⓔ   substitution property of equality

2. *Multiple Choice*   The basic axiom of algebra represented by $7(x + y) = 7x + 7y$, where x and y are real numbers, is ___?___.
   Ⓐ   closure property of multiplication
   Ⓑ   commutative property of multiplication
   Ⓒ   associative property of multiplication
   Ⓓ   distributive property of multiplication
   Ⓔ   associative property of addition

3. *Multiple Choice*   The basic axiom of algebra represented by $(12x)y = 12(xy)$, where x and y are real numbers, is ___?___.
   Ⓐ   closure property of multiplication
   Ⓑ   commutative property of multiplication
   Ⓒ   associative property of multiplication
   Ⓓ   distributive property of multiplication
   Ⓔ   identity property of multiplication

4. *Multiple Choice*   The basic axiom of algebra represented by $(1 + 2) + 3 = 1 + (2 + 3)$ is ___?___.
   Ⓐ   closure property of addition
   Ⓑ   inverse property of addition
   Ⓒ   identity property of addition
   Ⓓ   commutative property of addition
   Ⓔ   associative property of addition

Multiple Choice   In Exercises 5–10, list the correct axiom for the problem below using the choices A–E. You may use a choice more than once.

   Ⓐ   Definition of subtraction
   Ⓑ   Distributive property of multiplication
   Ⓒ   Distributive property of addition
   Ⓓ   Associative property of multiplication
   Ⓔ   Substitution property of equality

For all real numbers x and y, and where $z = x$, then $5(x - y) - z = 4x - 5y$.

Prove: $5(x - y) - z = 4x - 5y$

   5. $5(x - y) - z = 5(x - y) - x$
   6. $= 5[x + (-y)] - x$
   7. $= 5x + 5(-y) - x$
   8. $= 5x + (-5y) - x$
   9. $= 5x - 5y - x$
   10. $= 4x - 5y$

11. *Multi-Step Problem*   Use the basic axioms of algebra to prove that if $2x + 5 = 11$, then $x = 3$. Use the two column format shown.

Statement	**Reason**
1. $2x + 5 = 11$	Given
2. $2x + 5 + (-5) = 11 + (-5)$	_____
3. $2x + 0 = 11 + (-5)$	_____
4. $2x = 11 + (-5)$	_____
5. $2x = 6$	_____
6. $\frac{1}{2} \cdot 2x = 6 \cdot \frac{1}{2}$	_____
7. $1(x) = 6 \cdot \frac{1}{2}$	_____
8. $x = 6 \cdot \frac{1}{2}$	_____
9. $x = 3$	_____

Chapter 12

Cumulative Standardized Test Practice

For use after Chapters 1–12

1. *Multiple Choice*   Evaluate the expression $x^2 - 3y^2 + 8$ when $x = 2$ and $y = -2$.

   Ⓐ 12   Ⓑ 0   Ⓒ -12

   Ⓓ 24   Ⓔ -24

2. *Multiple Choice*   Simplify $3x(x - 2) - 2x + x^2$.

   Ⓐ $2x^2 - 8x$ Ⓑ $4x^2 - 4x$

   Ⓒ $4x^2 - 2x - 2$ Ⓓ $4x^2 + 8x$

   Ⓔ $4x^2 - 8x$

Quantitative Comparison   In Exercises 3–6, choose the statement below that is true about the given quantities.

   Ⓐ The value in column A is greater.

   Ⓑ The value in column B is greater.

   Ⓒ The two values are equal.

   Ⓓ The relationship cannot be determined from the given information.

	Column A	Column B
3.	$12x - 3 = 4$	$2x - 5 = 3x - 8$
4.	$\frac{1}{2}x + |-3| = 7$	$7x + 2(-x + 6) = 3$
5.	The slope of $y = \frac{1}{2}x - \frac{1}{2}$	The y-intercept of $y = \frac{1}{2}x - \frac{1}{2}$
6.	$-15 = \frac{x}{5}$	$\frac{1}{5}x = -15$

7. *Multiple Choice*   There are 5 yellow golf balls, 3 red golf balls, and 10 white golf balls in a bag. What are the odds of randomly choosing a yellow golf ball?

   Ⓐ $\frac{5}{13}$ Ⓑ $\frac{5}{18}$ Ⓒ $\frac{3}{18}$

   Ⓓ $\frac{13}{5}$ Ⓔ $\frac{18}{5}$

8. *Multiple Choice*   What is the length of a rectangular container if its volume is 336 cubic inches? Its height is 6 inches and its width is 8 inches. (*Hint*: $V = lwh$)

   Ⓐ 6 in.   Ⓑ 8 in.   Ⓒ 5 in.

   Ⓓ 7 in.   Ⓔ 3 in.

9. *Multiple Choice*   Find the slope of the line passing through the points $(5, -2)$ and $(11, 2)$.

   Ⓐ 0   Ⓑ $\frac{2}{3}$ Ⓒ $-\frac{2}{3}$

   Ⓓ $\frac{3}{2}$ Ⓔ $-\frac{3}{2}$

10. *Multiple Choice*   What is an equation of a line perpendicular to $y = \frac{1}{3}x - 5$?

   Ⓐ $y = 3x + 6$ Ⓑ $y = \frac{1}{3}x + \frac{1}{5}$

   Ⓒ $y = -\frac{1}{3}x + 2$ Ⓓ $y = -3x - 8$

   Ⓔ $y = -\frac{1}{3}x + 5$

11. *Multiple Choice*   What is the standard form of an equation that passes through the points $(14, 8)$ and $(3, 11)$?

   Ⓐ $11y + 3x = 118$

   Ⓑ $11y - 3x = 130$

   Ⓒ $11y + 3x = -130$

   Ⓓ $11y - 3x = -118$

   Ⓔ $11y + 3x = 130$

12. *Multiple Choice*   What is the mean of the set of numbers: 5, 8, 11, 3, 14, 1, 2?

   Ⓐ 6.3   Ⓑ 5.8   Ⓒ 7.3

   Ⓓ 8.1   Ⓔ 9.3

13. *Multiple Choice*   What is the solution of the inequality $8 \leq -3x + 1 \leq 11$?

   Ⓐ $-3 \leq x \leq -4$ Ⓑ $-3 \geq x \geq -4$

   Ⓒ $3 \geq x \geq 4$ Ⓓ $-\frac{10}{3} \geq x \geq -\frac{7}{3}$

   Ⓔ $-\frac{10}{3} \leq x \leq -\frac{7}{3}$

14. *Multiple Choice*   If $y = 4x - 2$ and $7x - 2y = 1$ then $x = $ _____?_____ .

   Ⓐ 1   Ⓑ 3   Ⓒ 10

   Ⓓ 0   Ⓔ 5

15. *Multiple Choice*   Use the substitution method to solve the system of linear equations.

$$3x + 4y = 14$$
$$y = -\tfrac{5}{2}x$$

A $(-1, 8)$ **B** $(4, -10)$

C $(-2, 5)$ **D** $(2, 2)$ **E** $\left(\tfrac{2}{3}, 3\right)$

16. *Multiple Choice*   Solve the system of linear equations using linear combinations.

$$5x - 3y = 26$$
$$3x + 2y = 8$$

A $(-2, 4)$ **B** $(7, 3)$ **C** $(2, 4)$

D $(4, -2)$ **E** $(0, 4)$

17. *Multiple Choice*   Choose the system of linear inequalities represented by the graph.

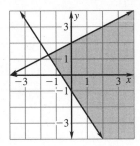

A $y \le \tfrac{1}{2}x + 2$ **B** $y \le \tfrac{1}{2}x + 2$

   $y \ge -\tfrac{3}{2}x - 1$ $y \le -\tfrac{3}{2}x - 1$

C $y \ge \tfrac{1}{2}x + 2$ **D** $y \ge \tfrac{1}{2}x + 2$

   $y \ge -\tfrac{3}{2}x - 1$ $y \le -\tfrac{3}{2}x - 1$

E   None of the above

18. *Multiple Choice*   Simplify the equation

$$\frac{5x^2y^{-3}}{8x^3y} \cdot \frac{4x^{-2}y^3}{10xy^{-4}}.$$

A $\dfrac{x^2y^3}{4}$ **B** $\dfrac{y^3}{4x^2}$ **C** $\dfrac{y}{4x^4}$

D $4y^3x^4$ **E** $\dfrac{y^3}{4x^4}$

Quantitative Comparison   In Exercises 19–21, choose the statement that is true about the given quantity.

A   The value of column A is greater.

B   The value of column B is greater.

C   The two values are equal.

D   The relationship cannot be determined from the information given.

	Column A	Column B
19.	$(2x)^2(x^2)^3$ when $x = 2$	$(7x^3)^2$ when $x = 2$
20.	$\left(\tfrac{3}{2}\right)^{-3}$	$\left(\tfrac{1}{2}\right)^2\left(\tfrac{1}{3}\right)^{-2}$
21.	$y = 14\left(\tfrac{9}{7}\right)^t$ when $t = 2$	$y = 6(1.5)^t$ when $t = 2$

22. *Multiple Choice*   You deposit $1200 into a savings account that pays 7% interest compounded yearly. How much money is in the account after 8 years assuming you made no additional investments or withdrawls? (*Hint: I = Prt*)

A $1872   **B** $2056.59

C $2117.86   **D** $2061.82

E $8370.91

23. *Multiple Choice*   Which is the simplified form of $\dfrac{2\sqrt{45} \cdot \sqrt{15}}{-4\sqrt{6}}$?

A $-5\sqrt{12}$ **B** $\dfrac{-\sqrt{10}}{2}$

C $\dfrac{-15\sqrt{2}}{4}$ **D** $\dfrac{-12\sqrt{3}}{24}$

E $\dfrac{-20\sqrt{2}}{4}$

Chapter 12

24. Multiple Choice Evaluate $\sqrt{b^2 - 4ac}$ when $a = 7, b = -2, c = -6$.

(A) $2\sqrt{51}$ (B) $\sqrt{97}$

(C) $2\sqrt{43}$ (D) $2\sqrt{23}$

(E) undefined

25. Multiple Choice What is the vertex of the graph of the equation $y = x^2 + 4x - 12$?

(A) $(-4, -18)$ (B) $(-2, -16)$

(C) $(0, -12)$ (D) $(12, -2)$

(E) $(2, 0)$

26. Multiple Choice What are the solutions of $y = x^2 - 5x - 50$?

(A) $-25, 2$ (B) $2, 25$

(C) $-5, -10$ (D) $5, -10$

(E) $-5, 10$

27. Multiple Choice Use the quadratic formula to find a solution for $-2x^2 - 3x + 9 = 0$.

(A) 3   (B) 6   (C) -6

(D) $-\frac{3}{2}$ (E) $\frac{3}{2}$

28. Multiple Choice Use the discriminant to determine the number of solutions of the equation $3x^2 - 2x + 8 = 0$.

(A) 1   (B) 2   (C) 3

(D) Infinitely Many   (E) None

29. Multiple Choice Which ordered pair is a solution of the inequality $y > 2x^2 - 9x - 18$?

(A) $(-1, -9)$ (B) $(-3, 17)$ (C) $(2, 6)$

(D) $(0, -22)$ (E) $(-2, 4)$

30. Multiple Choice Which of the following is equal to $(x + 7)(x - 3)$?

(A) $x^2 + 4x - 21$ (B) $x^2 + 10 - 21$

(C) $x^2 + 10x - 21$ (D) $x^2 - 10x + 4$

(E) $x^2 - 4x - 21$

Quantitative Comparison In Exercises 31 and 32, add the solutions of the equations. Choose the statement below that is true about the given number.

(A) The solution of column A is greater.

(B) The solution of column B is greater.

(C) The solutions are equal.

(D) The relationship cannot be determined from the information given.

	Column A	Column B
31.	$x^2 - 8x - 24 = 20$	$x^2 + 24 = -10x$
32.	$x^2 - \frac{2}{3}x + \frac{1}{9} = 0$	$9x^2 - 36 = 0$

33. Multiple Choice Which of the following is the complete factorization of $6x^3 + 39x^2 - 21x$?

(A) $(3x^2 + 7)(2x - 3)$

(B) $3x(x + 7)(2x - 1)$

(C) $3x(2x + 7)(x - 1)$

(D) $3x(2x + 1)(x - 7)$

(E) $(3x + 3)(2x^2 - 7x)$

Quantitative Comparison In Exercises 34–36, choose the statement below that is true about the given number.

(A) The value in column A is greater.

(B) The value in column B is greater.

(C) The two values are equal.

(D) The relationship cannot be determined from the information given.

	Column A	Column B
34.	$\dfrac{3x}{5} = \dfrac{17}{4}$	$\dfrac{5}{9} = \dfrac{2x}{25}$
35.	15% of 120 is x	85 of 290% is x
36.	$\dfrac{x + 2}{18} = \dfrac{x}{8}$	$\dfrac{2x - 3}{15} = \dfrac{x + 6}{12}$

37. *Multiple Choice* The variables x and y vary inversely. When x is 10, y is 7. If x is 5, then y is ___?___ .

- **(A)** 21
- **(B)** 3.5
- **(C)** 350
- **(D)** -7
- **(E)** 14

38. *Multiple Choice* Simplify the expression.

$$\left(\frac{30x^2}{x+2} + \frac{2x}{x^2}\right) \div \frac{15x^3 - 30x^2}{x^2 - 4}$$

- **(A)** $\dfrac{30x^2 + 2x + 2}{x(x+2)^2}$
- **(B)** $\dfrac{15x^2 + x + 1}{15x^3}$
- **(C)** $\dfrac{30x^3 + 2x + 4}{15x^3}$
- **(D)** $\dfrac{x+1}{x^2}$
- **(E)** $\dfrac{30x^2 + 2x - 4}{x(x+2)(x-2)}$

39. *Multiple Choice* What is $14x^2 + 12x - 6$ divided by $7x - 1$?

- **(A)** $2x + 1 + \dfrac{3x - 6}{7x - 1}$
- **(B)** $2x + 2 + \dfrac{-8}{7x - 1}$
- **(C)** $2x + 2 + \dfrac{-4}{7x - 1}$
- **(D)** $7x + 1$
- **(E)** $7x + 2 + \dfrac{5x - 4}{7x - 1}$

40. *Multiple Choice* What is the domain of the function $y = 6 - \sqrt{x - 3}$?

- **(A)** $x \geq 6$
- **(B)** $x \leq 6$
- **(C)** $x \geq -3$
- **(D)** $x \geq 3$
- **(E)** $x \leq 3$

41. *Multiple Choice* The geometric mean of x and 4 is 12. What is the value of x?

- **(A)** 36
- **(B)** 38
- **(C)** 7
- **(D)** 3
- **(E)** 9

42. *Multiple Choice* Simplify $\dfrac{16}{2 - \sqrt{3}}$

- **(A)** $32 + 16\sqrt{3}$
- **(B)** $\dfrac{32 + 16\sqrt{3}}{11}$
- **(C)** $\dfrac{18\sqrt{3} + 3}{11}$
- **(D)** $32 - 16\sqrt{3}$
- **(E)** $18 + \sqrt{3}$

43. *Multiple Choice* What term should be added to $x^2 - \frac{2}{3}x$ so that the result is a perfect square trinomial?

- **(A)** $\frac{4}{9}$
- **(B)** $\frac{4}{3}$
- **(C)** $-\frac{1}{9}$
- **(D)** $\frac{1}{9}$
- **(E)** $\frac{1}{3}$

44. *Multiple Choice* What is the length of the missing side of the triangle?

- **(A)** 4
- **(B)** 8
- **(C)** 16
- **(D)** 5
- **(E)** 24

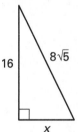

45. *Multiple Choice* What is the distance between $(7, -3)$ and $(-1, 4)$?

- **(A)** 6.1
- **(B)** 9.2
- **(C)** 8.1
- **(D)** 9.5
- **(E)** 10.6

46. *Multiple Choice* What is the length of side a in the triangle?

- **(A)** 45
- **(B)** 6
- **(C)** 7
- **(D)** 8
- **(E)** 9

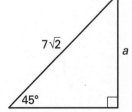

Chapter 12

ANSWERS

Chapter 1

Lesson 1.1 **1.** D **2.** C **3.** A **4.** E **5.** E
6. C **7.** A **8.** B **9.** B **10.** B **11.** D
12. A **13.** B

Lesson 1.2 **1.** E **2.** B **3.** B **4.** A **5.** E
6. D **7.** D **8.** C **9.** C **10. a.** 12 ft³
b. 89.76 gallons **c.** 67.32 gallons **d.** Yes

Lesson 1.3 **1.** E **2.** D **3.** C **4.** A **5.** E
6. B **7.** B **8.** B **9.** E **10.** A **11.** C
12. B

Lesson 1.4 **1.** B **2.** C **3.** B **4.** A **5.** D
6. C **7.** E **8.** A **9.** D **10.** B
11. a. $20w + 150 = 500$ **b.** 150 is constant
savings **c.** 17.5 weeks **d.** 23.3 weeks

Lesson 1.5 **1.** D **2.** B **3.** B **4.** A **5.** E
6. C **7.** E **8.** C **9.** D **10.** B **11.** A

Lesson 1.6 **1.** A **2.** D **3.** C **4.** C **5.** E
6. B **7.** A **8.** C

Lesson 1.7 **1.** A **2.** C **3.** C **4.** E **5.** B
6. a. $1.25x = p$ **b.** $1.25x - 75 = p$ **c.** 860
d. 950

Chapter 2

Lesson 2.1 **1.** B **2.** D **3.** E **4.** E **5.** A
6. E **7.** A **8.** D **9.** C **10.** B **11.** C
12. D

Lesson 2.2 **1.** C **2.** D **3.** C **4.** A **5.** C
6. D **7.** E **8.** D **9.** B **10.** B **11.** A
12. C

Lesson 2.3 **1.** B **2.** C **3.** D **4.** B **5.** D
6. D **7.** A **8.** E **9.** B **10.** C
11. a. $-2, 11, -2, -11$ **b.** a decrease in birds
spotted **c.** an increase in birds spotted

Lesson 2.4 **1.** C **2.** B **3.** C **4.** A **5.** C
6. A **7.** B **8.** A **9.** A

Lesson 2.5 **1.** A **2.** D **3.** B **4.** A **5.** A
6. A **7.** C **8.** E **9.** C
10. a. $75 - 3h = m$ **b.** $39 **c.** 25 hours

Lesson 2.6 **1.** B **2.** A **3.** E **4.** D **5.** D
6. C **7.** B **8.** B **9. a.** C **b.** 5.2 hours

Lesson 2.7 **1.** B **2.** C **3.** A **4.** B **5.** D
6. D **7.** A **8.** B **9.** A **10.** B **11.** A
12. A **13.** A

Lesson 2.8 **1.** C **2.** B **3.** C **4.** D **5.** A
6. C **7.** E **8.** C **9.** B **10.** B

Chapter 3

Lesson 3.1 **1.** B **2.** B **3.** C **4.** D **5.** A
6. E **7.** D **8.** C **9.** A **10.** B **11.** A
12. B **13.** D **14.** A

Lesson 3.2 **1.** C **2.** D **3.** C **4.** A **5.** A
6. B **7.** E **8.** D **9.** A **10.** C **11.** A

Lesson 3.3 **1.** B **2.** C **3.** A **4.** A **5.** D
6. E **7.** D **8.** B **9.** C **10.** C **11.** B
12. A **13.** D

Lesson 3.4 **1.** C **2.** B **3.** A **4.** E **5.** B
6. B **7.** C **8.** D **9.** D **10.** A **11.** C
12. A

Lesson 3.5 **1.** A **2.** B **3.** C **4.** E **5.** D
6. D **7. a.**

b. 6.2 mi/hr **c.** 20 miles **d.** 6.7 mi/hr

Lesson 3.6 **1.** C **2.** E **3.** E **4.** D **5.** B
6. A **7.** B **8.** B **9.** C

Chapter 3 *continued*

10. a. $29.95 + 0.32(x - 30) = 46.82$ **b.** 82.7
c. 52.7

Lesson 3.7 **1.** E **2.** B **3.** A **4.** C **5.** C
6. E **7.** D **8.** B **9.** B **10.** C **11.** A

Lesson 3.8 **1.** D **2.** B **3.** B **4.** C **5.** D
6. A **7.** D **8.** E

9. a. $\dfrac{\text{Brown in Sample}}{\text{Total Sample}} = \dfrac{\text{Brown in Ship.}}{\text{Total Ship.}}$

b. 40   **c.** reject

Chapter 4
Lesson 4.1 **1.** B **2.** A **3.** E **4.** C **5.** B
6. D **7.** B

8.
a.
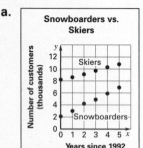
b. slightly increasing
c. rapidly increasing

Lesson 4.2 **1.** D **2.** B **3.** C **4.** A **5.** C
6. C **7.** A **8.** E **9. a.** $y = -2x + 36$
b.

x	5	10	15
y	26	16	6

c.
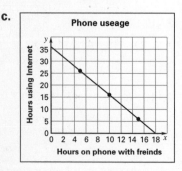
d. 28

Lesson 4.3 **1.** B **2.** E **3.** E **4.** C **5.** B
6. E **7.** B **8.** D **9.** B **10.** A

Lesson 4.4 **1.** C **2.** D **3.** A **4.** B **5.** A
6. E **7.** C **8.** B **9.** C **10.** B **11.** A

12. D

Lesson 4.5 **1.** A **2.** B **3.** D **4.** E **5.** D
6. C **7.** C **8.** B **9.** A **10.** C

Lesson 4.6 **1.** C **2.** B **3.** C **4.** D **5.** C
6. D **7.** C **8.** E **9.** B **10.** C **11.** A

Lesson 4.7 **1.** D **2.** E **3.** D **4.** C **5.** B
6. A **7. a.** Model 1:
$350,000 = 13,850x + 220,000$
b. Model 2: $350,000 = 24,750x + 220,000$

Lesson 4.8 **1.** C **2.** A **3.** D **4.** D **5.** E
6. B **7.** C **8.** B **9.** A **10.** A

Chapter 5
Lesson 5.1 **1.** B **2.** E **3.** A **4.** C **5.** B
6. B **7.** B **8.** C **9.** A

Lesson 5.2 **1.** C **2.** B **3.** D **4.** C **5.** B
6. E **7.** A **8.** B **9.** A

Lesson 5.3 **1.** D **2.** A **3.** C **4.** E **5.** B
6. C **7.** B
8. a. $y = \frac{5}{2}x + \frac{11}{2}; y = -\frac{2}{5}x + \frac{13}{5}; y = \frac{3}{7}x - \frac{5}{7}$
b. $\overline{AB}$ and $\overline{BC}$; perpendicular slopes

Lesson 5.4 **1.** E **2.** C **3.** B **4.** A **5.** D
6. C

7.
a.
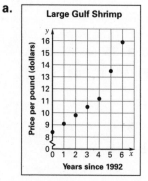
b. Answers may vary.
Sample answer:
$y = \frac{6}{5}x + 8$
c. about $18.80 per pound

Lesson 5.5 **1.** D **2.** B **3.** A **4.** B **5.** E
6. C **7.** B **8.** A **9.** B

Lesson 5.6 **1.** D **2.** E **3.** B **4.** C **5.** A
6. C **7.** B **8.** A **9.** B **10.** D

Algebra 1
Standardized Test Practice Workbook

Chapter 5 *continued*

Lesson 5.7
1. D 2. B 3. C 4. D

5. a.

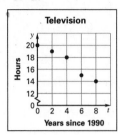

b.

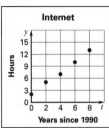

c. $y = -\frac{3}{4}t + 20$; $y = \frac{5}{4}t + 2$ d. 15.75 hrs; extrapolation e. 16.25 hrs; interpolation

Chapter 6

Lesson 6.1
1. B 2. C 3. C 4. C 5. D
6. E 7. A 8. B 9. A 10. C

Lesson 6.2
1. B 2. A 3. A 4. D 5. C
6. C 7. a. $x \geq 660$ b. $x \geq 165$ c. $x \geq 7$

Lesson 6.3
1. A 2. D 3. C 4. D 5. B
6. B 7. A 8. D 9. C

Lesson 6.4
1. B 2. D 3. A 4. A 5. E
6. C 7. D 8. B 9. A 10. C

Lesson 6.5
1. D 2. A 3. B 4. C 5. B
6. E 7. a. $8.5x + 10.5y \leq 485$

b.

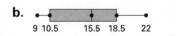

c. 30 or less

d. No, you can't order fractional parts of a meal.

Lesson 6.6
1. C 2. E 3. C 4. C 5. D
6. E 7. B 8. A 9. C 10. B 11. D

Lesson 6.7
1. B 2. A 3. A 4. C 5. E
6. A 7. C 8. a.

c. 22; 15.5

b.

d. Juice purchases average between 16 and 27 units, and milk between 9 and 22. The range for juice is 11, and for milk it is 13. Juice sold more than 19 units 75% of days while milk sold less than 18.5 units 75% of the days.

Cumulative Review

Chapters 1–6
1. B 2. C 3. A 4. E 5. B
6. A 7. A 8. D 9. C 10. D 11. B
12. E 13. D 14. C 15. C 16. B 17. C
18. B 19. A 20. A 21. E
22. a. $32 + 12(x - 1)$ b. 3 c. Divide the total by 1.06 to get the price before tax. 23. D
24. E 25. A 26. B 27. D 28. D 29. B
30. C 31. D 32. E 33. B 34. A 35. C
36. A 37. A 38. D 39. B 40. B

41.
a.

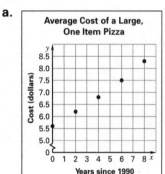

b. $y = \frac{1}{3}x + 5.60$
c. $7.27; interpolation
d. $9.60

42. B 43. C 44. B 45. A 46. C 47. D
48. E 49. A

Chapter 7

Lesson 7.1
1. E 2. C 3. B 4. C 5. D
6. D 7. B 8. A 9. D 10. A

Lesson 7.2
1. B 2. C 3. E 4. B 5. A
6. D 7. C 8. A 9. C 10. B

Lesson 7.3
1. B 2. D 3. E 4. C 5. A
6. a. $5 + c = 6$ b. 5 mi/h c. 1 mi/h
$5 - c = 4$

d. No, not without additional information.

Lesson 7.4
1. B 2. C 3. C 4. E 5. D
6. D 7. A 8. C 9. B

Chapter 7 *continued*

Lesson 7.5 **1.** D **2.** B **3.** A **4.** B **5.** D
6. B **7.** A **8.** C **9.** B

Lesson 7.6 **1.** A **2.** D **3.** B **4.** D **5.** E
6. a. $x + y \le 25$ **b.**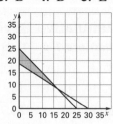
$\quad\quad 5x + 8y \ge 150$
$\quad\quad x \ge 0$
$\quad\quad y \ge 0$

c. Answers may vary. *Sample answer*: 15 hours
at first job and 10 jours at second job.

Chapter 8

Lesson 8.1 **1.** B **2.** C **3.** D **4.** E **5.** A
6. B **7.** C **8.** D **9.** D **10.** C **11.** B
12. A

Lesson 8.2 **1.** C **2.** B **3.** B **4.** A **5.** E
6. B **7.** D **8.** D **9.** D **10. a.** $367.51
b. $500 **c.** $583.20

Lesson 8.3 **1.** B **2.** E **3.** C **4.** D **5.** A
6. E **7.** D **8.** D **9.** E **10.** B **11.** A
12. C **13.** A

Lesson 8.4 **1.** A **2.** D **3.** B **4.** D **5.** E
6. C **7.** B **8.** C **9.** A **10.** A **11.** B
12. B

Lesson 8.5 **1.** C **2.** D **3.** B **4.** B **5.** C
6. D **7. a.** $y = 50{,}000(1.15)^t$ **b.** 133,000
c. 37,807 **d.** Answers may vary. *Sample
answer:* It will probably continue to increase, but
not as rapidly.

Lesson 8.6 **1.** E **2.** C **3.** C **4.** A **5.** C
6. D **7.** A **8.** C **9.** A

Chapter 9

Lesson 9.1 **1.** D **2.** E **3.** D **4.** D **5.** E
6. A **7.** B **8.** C **9.** A **10.** C **11.** D
12. B **13.** B

Lesson 9.2 **1.** D **2.** C **3.** E **4.** A **5.** D
6. B **7.** D **8.** D **9.** B **10.** C **11.** A

Lesson 9.3 **1.** D **2.** B **3.** B **4.** C **5.** E
6. C **7.** D **8.** A **9.** A **10.** C **11.** B

Lesson 9.4 **1.** D **2.** A **3.** C **4.** B **5.** B
6. A **7.** C **8.** A **9.** C **10.** E **11.** D

Lesson 9.5 **1.** B **2.** E **3.** B **4.** C **5.** B
6. A **7.** E **8.** B **9. a.** $n = -16t^2 + 50$
b. about 1.77 sec **c.** *Sample answer*: The speed
of the ski lift if it is moving when the keys are
dropped, the angle that the keys are dropped, and
the wind and weather conditions.

Lesson 9.6 **1.** B **2.** E **3.** C **4.** C **5.** B
6. C **7.** C **8.** A **9.** C **10.** B

Lesson 9.7 **1.** B **2.** E **3.** A **4.** D **5.** A
6. a. 4 feet **b.** 4 feet **c.** Not if it layed
outside of $y \le x^2 - 2x - 3$.

Lesson 9.8 **1.** A **2.** C **3.** B **4.** B **5.** D
6. E **7.** D **8.** A **9.** C
10. a. 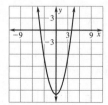 **b.** quadratic
c. $y = x^2 - 16$

Chapter 10

Lesson 10.1 **1.** E **2.** A **3.** D **4.** B **5.** B
6. B **7. a.** $S = 8.9t^2 + 1.2t + 73$
b. $111,000 **c.** $350 = 21.t^2 + 7.5t + 138$

Lesson 10.2 **1.** D **2.** E **3.** A **4.** E **5.** D
6. C **7.** A **8.** B **9.** B **10.** A

Lesson 10.3 **1.** E **2.** B **3.** D **4.** E **5.** E
6. B **7.** A **8.** A **9.** B

Lesson 10.4 **1.** E **2.** B **3.** A **4.** C **5.** C
6. D **7.** C **8. a.** 12 feet **b.** 6.5 feet
c. 6 people

Chapter 10 *continued*

Lesson 10.5 **1.** B **2.** C **3.** E **4.** C **5.** A
6. D **7.** D **8.** C **9.** A **10.** A **11.** D
12. A

Lesson 10.6 **1.** B **2.** D **3.** C **4.** B **5.** A
6. C **7.** E **8.** D
9. a. $(-4t + 5)(4t + 4) = 0$ **b.** $\frac{5}{4}$ and -1
c. The correct answer is 1.25 seconds. Time
cannot be negative for this problem.

Lesson 10.7 **1.** D **2.** A **3.** B **4.** C **5.** D
6. D **7.** A **8.** E **9.** B **10.** B **11.** C

Lesson 10.8 **1.** B **2.** C **3.** B **4.** A **5.** C
6. D **7.** B **8.** E

Chapter 11
Lesson 11.1 **1.** E **2.** C **3.** A **4.** D **5.** B
6. C **7.** A **8.** D **9.** B **10.** A **11.** D

Lesson 11.2 **1.** B **2.** D **3.** A **4.** E **5.** B
6. C **7.** E **8.** E **9.** C **10.** A **11.** C
12. B **13.** C

Lesson 11.3 **1.** E **2.** A **3.** D **4.** D **5.** D
6. B **7.** C
8. a.

h	240	120	100
r	5.00	10.00	12.00

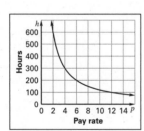

b. inverse; $y = \dfrac{1200}{x}$

c. The first bike is one
and a half times more
expensive than the
second. x and y would
relate similary.

Lesson 11.4 **1.** D **2.** B **3.** E **4.** D **5.** D
6. C **7.** B **8.** A **9.** A **10.** B **11.** D

Lesson 11.5 **1.** E **2.** B **3.** A **4.** B **5.** D
6. C **7. a.** $\dfrac{200(2000 + 23t)(1 + 0.02t)}{(1 - 0.03t)(4 + t)}$
b. about 56,700 **c.** about 54,900

d. The miles driven per person is decreasing; this
can't continue indefintely.

Lesson 11.6 **1.** D **2.** E **3.** D **4.** B **5.** A
6. a. $\dfrac{1200}{(x + 25)} + \dfrac{1200}{(x - 30)}$
b. $\dfrac{2400x - 6000}{(x + 25)(x - 30)}$ **c.** about 12.4 hours
d. The smaller plane is traveling at half the speed
of the larger plane. At first glance, the trip might
seem to be exactly twice as long, but due to wind
force the trip actually takes about 26.7 hours.

Lesson 11.7 **1.** C **2.** B **3.** D **4.** D **5.** B
6. B **7.** D **8.** B **9.** E

Lesson 11.8 **1.** B **2.** B **3.** A **4.** C **5.** E
6. B **7.** E **8.** A **9.** B **10.** B

Chapter 12
Lesson 12.1 **1.** A **2.** A **3.** B **4.** A **5.** C
6. C **7.** C **8.** B **9.** B

Lesson 12.2 **1.** D **2.** A **3.** E **4.** B **5.** C
6. E **7.** B **8.** B **9.** C

Lesson 12.3 **1.** D **2.** C **3.** B **4.** A **5.** B
6. C **7.** E **8.** D **9.** A **10.** C **11.** D
12. B **13.** A

Lesson 12.4 **1.** D **2.** C **3.** B **4.** B **5.** C
6. B **7.** D **8.** E **9.** C **10.** A **11.** B

Lesson 12.5 **1.** C **2.** A **3.** B **4.** B **5.** D
6. E **7.** D **8.** D **9. a.** $x, 2x - 30$
b.

2x − 30, x, 200

c. 135.9 mi/h, 241.8 mi/h **d.** Answers may vary.
Sample Answer: The quadratic formula was used.
Although this is factorable, there are many
combinations to check.

Lesson 12.6 **1.** D **2.** B **3.** A **4.** E **5.** C
6. C **7.** B **8.** D **9.** D **10.** D **11.** A
12. C **13.** A

Answers

Lesson 12.7 **1.** C **2.** B **3.** E **4.** D **5.** A
6. E **7.** E

8. a.

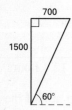

b. about 1655

c. No, the triangles will not always be right triangles.

Lesson 12.8 **1.** D **2.** D **3.** C **4.** E **5.** E
6. A **7.** B **8.** D **9.** A **10.** A

11. Reason

1. Given

2. Addition Property of Equality

3. Inverse Property of Addition

4. Identity Property of Addition

5. Substitution

6. Multiplication Property of Equality

7. Inverse Property of multiplication

8. Identity Property of multiplication

9. Substitution

Cummulative Review

Chapters 1–12 **1.** B **2.** E **3.** B **4.** A
5. A **6.** C **7.** A **8.** D **9.** B **10.** D
11. E **12.** A **13.** E **14.** B **15.** C **16.** D
17. A **18.** E **19.** B **20.** B **21.** A **22.** D
23. C **24.** C **25.** B **26.** E **27.** E **28.** E
29. C **30.** A **31.** A **32.** A **33.** B **34.** A
35. B **36.** B **37.** E **38.** C **39.** C **40.** D
41. A **42.** A **43.** D **44.** B **45.** E **46.** C